ABOUT THE A[UTHOR]

Dr Stephen Juan, "the Wizard of Odds", is the author of the "Odd Books" series. The books in the series are *The Odd Body* (1995), *The Odd Brain* (1998), *The Odd Body 2* (2000), *The Odd Sex* (2001) and now *The Odd Body 3* (2007).

Scientist, educator and journalist, Dr Juan is an anthropologist by training and one of the world's foremost communicators of research. His various books have been translated into 22 languages. He has served as a magazine editor and a newspaper columnist for papers such as the *Sydney Morning Herald*, the *Sun-Herald* (Sydney), the *National Post* (Toronto), the *New York Daily News* and the *Register* (London). Lively and popular as a speaker, Dr Juan also appears regularly on news and current affairs television and radio programs covering any and all topics to do with being a human being.

Born in California, Dr Juan received his university training at the University of California at Berkeley. For 29 years, Dr Juan has taught in the Faculty of Education and Social Work at the University of Sydney. Dr Juan has been honoured for his writing and other public education work, by the American Medical Association and the US National Association of Physician Broadcasters among others.

The Odd Body 3 is Dr Stephen Juan's eleventh book.

THE ODD BODY 3

STILL MORE MYSTERIES OF OUR WEIRD AND WONDERFUL BODIES EXPLAINED

THE ODD BODY

3

STILL MORE MYSTERIES OF OUR WEIRD AND WONDERFUL BODIES EXPLAINED

DR STEPHEN JUAN

HarperCollinsPublishers

To all the wonderful, life-saving staff of the Royal Prince Alfred
Hospital in Sydney, where many angels daily tread

HarperCollins*Publishers*

First published in Australia in 2007
by HarperCollins*Publishers* Australia Pty Limited
ABN 36 009 913 517
www.harpercollins.com.au

Text copyright © Stephen Juan and Associates 2007

The right of Stephen Juan to be identified as the author of this work has been
asserted by him under the Copyright Amendment (Moral Rights) Act 2000.

HarperCollins*Publishers*
25 Ryde Road, Pymble, Sydney NSW 2073, Australia
31 View Road, Glenfield, Auckland 10, New Zealand
77–85 Fulham Palace Road, London W6 8JB, United Kingdom
2 Bloor Street East, 20th floor, Toronto, Ontario M4W 1A8, Canada
10 East 53rd Street, New York NY 10022, USA

National Library of Australia Cataloguing-in-Publication data:

Juan, Stephen.
    The odd body 3: still more mysteries of our weird and wonderful
    bodies explained.
    Bibliography references.
    Includes index.
    ISBN 13: 978 0 7322 8304 9
    ISBN 10: 0 7322 8304 3
    1. Body, Human – Miscellanea. I. Title.
612

Cover design by Darren Holt, HarperCollins Design Studio
Internal design adapted by Darren Holt, HarperCollins Design Studio
Typeset in 11/13 Bembo by Kirby Jones and Helen Beard, ECJ Australia Pty Ltd
Printed and bound in Australia by Griffin Press on 79gsm Bulky Paperback White

5 4 3 2 1   07 08 09 10

contents

INTRODUCTION

Hello, reader.

Welcome to *The Odd Body 3*. I'm so glad you have chosen to leaf through it, perhaps read a page or two of it, or maybe even read it from cover to cover. Whatever you do, I hope you enjoy it, learn from it, and are entertained by it.

Before *The Odd Body 3* came *The Odd Body* and *The Odd Body 2*. In a biblical-sounding sense, *The Odd Body* begat *The Odd Body 2*, which begat *The Odd Body 3*. In starting this series, no one ever thought there would be so much begetting. The fact that the third one has been begotten means that at least some people enjoy them. These books are now read all over the world — thank you, reader. It goes to show you that people everywhere are fascinated by the body. The five books in the "Odd Books" series prove there is a lot to be fascinated about.

It seems that questions about the human body are endless. They keep coming in faster than they can be answered. I have called them Odd Body Questions (OBQs for short) since the first book in the "Odd" series was published in Australia in 1995. I keep asking OBQs, mine and readers', and I keep searching for the answers. Although many OBQs from readers have been asked before, sometimes many times before (e.g., "Why do men have nipples?", "Why does hair turn grey?", "Why do we get wrinkles?"), sometimes there is a new OBQ that has never been asked, let alone answered. That's when I go to work — I usually drop everything to search for the answer. Off to the library, textbooks, journals, experts, the Internet and anywhere else to get the answer. Must find the answer by yesterday! I am often working on several OBQs at once. When I find the answer, I can't wait to get it into a newspaper column or into one of my books.

Readers have submitted some fantastic OBQs since *The Odd Body 2* was published; most are answered in this book. I didn't have space to get all the OBQs in, but I did my best. Readers of all ages have submitted OBQs and it's very interesting to see who asks which type of question. There is a pattern. Young readers usually, but by no means always, ask OBQs dealing with blood, gore, mucus, other body products. They also ask about puberty and about weird ways to die. Elderly readers ask OBQs dealing with maturity, illness and the ageing body's growing inefficiency. Readers in between ask OBQs about all of these. You would be surprised! One middle-aged woman asked many questions at great length about nose-picking. She even followed up her letter with a phone call. Often it is an urban myth that is being questioned and the question is often preceded by words to the effect of "Something I've been wondering about for years is ..."

There are over 800 OBQs in my files that have not yet been answered, and they are being added faster than they are being subtracted. I answer emails and letters personally, but do not always publish my answers. It's good to know that readers are just as curious to find the answer to their OBQ as I am. I am, after all, a rather "curious" person to begin with, and in more ways than 1!

In *The Odd Body 3* I have tried to span OBQs from head to toe, though I have left out a chapter on the brain that was in previous books. Perhaps soon I will write *The Odd Brain 2* and include the missing material. I have also left out a chapter on sex that was in *The Odd Body 2*. Perhaps soon I will write *The Odd Sex 2* and address those questions. So many books to write, so little time!

You will see that the notes and references for each chapter include many citations of my newspaper columns in the *Sydney Morning Herald*, the *Sun-Herald* (Sydney), the *National Post* (Toronto), the *New York Daily News* and the *Register* (London). This is so that readers who might have read this information before will have that reference. Writing a syndicated newspaper column each week is just as much fun as writing books — maybe even more! If your local newspaper does not run the column and you'd like to see it each week, let that newspaper's editor know.

But for now, here's *The Odd Body 3*. Have a good read!

CHAPTER 1

BEGINNINGS

In *Alice's Adventures in Wonderland*, Lewis Carroll wrote: "'Begin at the beginning,' the king said, gravely, 'and go on till you come to the end: then stop.'" As a wise person once said: "The beginning is easy; what happens next is much harder." There are so many questions to ask and answer about our beautiful, wonderful and fascinating bodies that it's hard to decide where to begin — or where to stop. But let's do what the king says and begin at the beginning. We'll look at birth, growth, human development and any other intriguing things about the beginnings of our odd body.

WHAT IS THE PROBABILITY THAT HUMAN LIFE EXISTS ON OTHER PLANETS?

Who knows? We can only guess. In 1961, the US astronomer Frank Drake proposed an equation for calculating the number of technologically advanced civilisations existing in our galaxy. The Drake equation can be expressed as: $N = R \times fp \times ne \times fl \times fi \times fc \times L$. "N" is the number of civilisations in the galaxy that have developed to the point of being able to communicate. "R" is the rate at which suitable stars are formed with the capability of forming planets such as ours. "fp" is the proportion of stars with planets. "ne" is the number of planets around any star with a temperature range that would be habitable by humans. "fl" is the proportion of planets on which life evolves. "fi" is the proportion of planets that reach the stage of human intelligence. "fc" is the proportion of planets that develop a communications technology similar to our own. "L" is the length of time for which an intelligent civilisation can hope to survive either accidental destruction by outside forces or self-destruction by misuse of its own technology.

Taking Drake's equation, a group of scientists called the Search for Extra-Terrestrial Intelligence (SETI) estimates the rate of star formation at about 20 per year (R = 20). SETI suggests that half of all stars will form planetary systems (fp = 0.5), that the number of planets within 1 system that could support life is 1 (ne = 1), and that life will appear and evolve on 1 in 5 such planets (fl = 0.2). Mindful that dolphins and whales are intelligent but have never developed technology, SETI also suggests that technology might be expected to appear in half the other worlds that support life (fc = 0.5). So putting these numbers into Drake's equation: N = 20 x 0.5 x 1 x 0.2 x 0.5 x L. The number of civilisations in the galaxy is equal to the number of years (L) that an advanced technological civilisation can hope to endure: N = L. Of course, the only such civilisation we have to go on is our own. It has only been technologically advanced for some 50 years. Thus, the number of advanced life forms in our galaxy is at least 50. But there are obviously many, many assumptions built into this, and we are changing our view of the galaxy itself based upon new research.[1]

WILL WE EVER DEVELOP AN ARTIFICIAL WOMB?

Yes, we will and in a sense we already have. Ectogenesis involves providing an apparatus that generates an artificial environment outside a woman's womb where a foetus can develop and grow until it is ready to breathe on its own. These devices are commonly called "artificial wombs". Some have already been built to be used with animals. Ignoring all ethical issues for the moment, artificial wombs may be developed within the next few years to the point where a third trimester (26 weeks gestation) or even a second trimester (13 weeks gestation) foetus could be transferred from its mother's uterus to an artificial womb. It could complete its growth inside this apparatus and be "born" at full term (38 to 40 weeks gestation). In theory, there appears to be no reason why an artificial womb cannot be developed that could accept a recently fertilised pre-embryo, support it for the full 9 months of gestation, and allow it to be removed when conditions are right for its "birth". All that seems to be required is advancement in this form of technology.

But who would want this for their child? First, a mother whose womb has a defect that will cause her baby to be born prematurely and face death or a life of disability would not have to worry about the prospect of either. Second, a mother who does not want to go through the risks, pain and discomforts of pregnancy, the dislocation of her career (including loss of income, job seniority etc), or whose own health is at risk could conceive a child, have the embryo transferred to an artificial womb, and return 9 months later to pick up her new baby. Dr Yoshinori Kuwabara of Juntendo University in Japan has developed a rectangular artificial womb made from acrylic and filled with heated amniotic fluid. The foetus lies submerged in the "tank womb", which replaces oxygen and cleans the foetal blood with a dialysis machine connected to the umbilical cord. The embryo is transferred to the "tank womb" at 3 weeks gestation and the machine takes it from there. Is this an answer for women who desperately want a child but often miscarry? Or is this the realisation of Aldous Huxley's nightmare in *Brave New World*?

WHAT CAN A BABY DO AT BIRTH?

(Asked by Dawn Williams of Penshurst, New South Wales)

Pregnancy is divided into 3 periods (trimesters) of roughly 3 months each. During the third trimester (26 to 40 weeks' gestational age), the human continues to grow steadily. At birth, the average baby weighs about 2.7 to 4 kilograms (6 to 8 pounds).

☞ The bones of the skull remain soft to make it easier for it to pass through the birth canal.

☞ The fine foetal hair (lanugo) has almost entirely disappeared by 38 to 40 weeks gestational age.

☞ The lungs have matured completely. Just before birth, a surface-active agent (surfactant) will coat the lungs, preventing them from being filled with water and allowing them to remain inflated.

☞ The baby's skin is covered with a protective creamy coating (vernix).

☞ The head will usually turn downward (lighten, or engage) by about 36 weeks gestational age. When this happens, check the fastest route to the hospital!

☞ For many babies, the irises of the eyes are slate blue. The permanent eye colour will not appear until several days or weeks after birth.

☞ The kidneys continue to mature.

☞ The brain continues to develop.

☞ The baby can suck its thumb.

☞ The vocal cords are capable of making crying sounds.

☞ The baby can see, especially things that are close up. Their best vision is at about 25 centimetres (10 inches).

☞ The baby hears well and does so even when sleeping.

☞ The baby smells well and recognises the smell of its own mother compared with the smells of other women.

☞ The baby has a preference for things that taste sweet.

☞ The baby dreams.

☞ The baby's skin is sensitive. Massaging a newborn has been shown to boost growth and development (see later in this chapter).

HOW OLD AND HOW BIG MUST AN INFANT BE IN ORDER TO SURVIVE?

(Asked by Dawn Williams of Penshurst, New South Wales)

A human should be born after a pregnancy lasting 40 weeks. By definition, any infant born prior to 20 weeks of gestation and weighing less than 500 grams is counted as a miscarriage. Such an infant cannot survive. It is too underdeveloped for survival in the environment outside the womb. Survival is defined as living for 28 days. It is rare, but possible, for an infant to survive at a gestational age of 21 weeks. A study conducted in The Netherlands lists the survival rates for extremely premature infants as:

22 weeks — 4.6 per cent
23 weeks — 46 per cent
24 weeks — 59 per cent
25 weeks — 82 per cent

As you can see, there's a big difference with each succeeding week.[2]

CAN WE TALK TO THE UNBORN?

Welcome to the wonderful world of haptonomy. If this word wasn't in your vocabulary, it is now. Research has established that we begin to communicate with other humans before we are born. Haptonomy is the scientific study of our ability to communicate at a deep, deep level. But unlike talking, haptonomic communication combines words, thoughts and feelings. It has been described as "a combination of listening and feeling", "spontaneous empathy", "psychotactile contact" or "the science of affectivity". But whatever the definition, it is a special sort of human-to-human relating. Clinical and developmental psychologist Dr Ludvig Janus of Heidelberg, Germany, cites research demonstrating that a communications link between the unborn child and other humans is very real indeed.[3] He shows how it is possible for parents to establish such a link and describes the results of establishing this link for both child and mother. Yet the major figure in this new field of research is Dr Frans Veldman.[4] In experiments conducted by Dr Veldman, it was observed under ultrasound imaging, that during the final trimester of pregnancy, if a father places his hand upon the bare abdomen of his pregnant partner an extraordinary thing happens: the unborn child "responds to the invitation to relate, moves itself toward his hand resting on the mother's belly and snuggles up into it". When the father's hand is taken away, the baby moves away.[5]

In any case, there is a distinct desire for the unborn baby to establish contact. Dr Janus claims that babies who have been born after engaging in haptonomy with parents "develop very well after their birth and that all measurements of their development distinguish them in a dramatic way from other babies". They develop higher IQs, better verbal skills, better perception and attention skills, less irritability, and fewer weight problems and gastrointestinal disturbances.

The benefits of haptonomy flow on to to the mother as well. For instance, many women have difficulty in labour due to the fact that the birth canal of the pelvis is simply too narrow. In particular, the final 2 centimetres of required space are normally missing in the human female. Drugs are often given to increase dilation. If there is no progress in this regard, eventually the baby and mother are at risk. This missing 2 centimetres often results in a difficult labour, a forceps delivery, or even a Caesarean delivery. But according to Dr Janus, it has been demonstrated that mothers who engage in haptonomy with their baby experience "a loosening up of the pubic cartilage and the sacropelvic joint. This in turn gives the birth canal an extra 2 centimetres of width — the crucial 2 centimetres that are normally missing in human females." Thus, the baby–mother communication works to their mutual benefit.

Of course, mothers have to be open to the possibility of this form of communication with their baby. A mother can "talk" to her baby in many ways. For example, she can wake up a sleeping baby inside her — without saying a word. This is frequently seen in hospitals during prenatal ultrasound exams: if a mother is told that there is no spontaneous movement of her baby observable on the monitor, and that this might be evidence of a disturbance to the pregnancy and to the baby, the mother is invariably shocked, especially when she cannot see the monitor. This shock to the mother is enough to wake up the baby. Intense baby movements almost always appear immediately on the monitor. This phenomenon was discovered in the 1980s by Dr Eric Reinhold, a neonatologist from Stuttgart, Germany. Dr Reinhold humorously observed, "Without saying a word, mother wakes you up. The first time now, but it probably won't be the last."[6] Sometimes it is almost as if the baby is desperately trying to "tell" the mother something. Instances abound of pregnant women who have to leave rock concerts because their baby refuses to calm down after exposure to the high-decibel noise levels. According to Dr B.R.H. Van den Bergh, a neonatologist at the University of Leuven in Belgium, this highlights the sensitivities of the unborn child — yet another aspect of haptonomy.[7]

Dr Janus observes that "a huge, unused, human potential lies buried in this area of prenatal relating. The infant mortality rate has been dramatically reduced in the last 100 years and now there is the chance that children come to the light of day not only physically healthy but also are given all that is required for good psychological development." Dr Veldman advises that parents take time to "establish affective links" with their unborn child. It is easy and natural, but it "sometimes takes time to learn to communicate with feeling". Dr Veldman writes that haptonomy research "demonstrates that faculties every human being should possess are nowadays more and more under-developed, they lie fallow, or are atrophied, if not totally absent. However, these faculties are of fundamental interest for contacts, interactions and human relationships." He laments that in our world, which is dominated by concerns of economic rationalism and the bottom line, we too often ignore the realm of "communication with feeling" — the affective side of our nature. "There is no room for feelings, emotions: for affectivity. The dimension of feeling — and all that concerns the affective life — is considered as lacking in interest and therefore has no place in this world, as it is of no economic or political value. This dimension would only disturb, in a awkward way, the economic processes of development and production." So if our babies are trying to tell us something, are we always willing to listen?[8, 9]

WHY DO WOMEN GIVING BIRTH HAVE LABOUR PAINS? WHY AREN'T THEY PROTECTED BY THE BODY'S ANAESTHETIC SYSTEM?

(Asked by Heather Andrews of St Kilda, Victoria)

As the joke goes, giving menstrual and labour pains to women and not men proves that God must be male. All kidding aside, the anaesthetic reactions in childbirth are different from the reaction to an actual physical injury. During physical trauma or physical injuries, the human body reacts using mechanisms that promote survival; in particular, the body has some immediate responses that allow an individual to use their cognitive functions without distraction from pain. This enables a person to avoid further traumatic injury by being able to escape from traumatic situations. Many times, a person with a quite severe physical injury is not mentally aware of it until much later.

In childbirth, the female body does not perceive the situation of uterine contractions and the stretching of the vaginal walls as a traumatic injury. Childbirth is an entirely natural and normal process; there is no sudden physical trauma, no actual injury incurred, and no real threat to the natural body systems. Childbirth involves contractions of the uterus that become increasingly stronger until the baby is born. And the contractions and other manifestations of the childbirth process are controlled by the body's own hormonal sequences. Thus, it is not a "traumatic" situation where something outside the body causes an injury.

Interestingly, in many if not most women, there is a rush of pain-killing endorphins during the childbirth experience. Many women tend to forget the actual pain over time. For instance, they might remember the hospital and the interactions with others, but the pain itself is not completely recalled. In addition, there are other methods that can be used to relieve pain in childbirth. Studies show that exercise can strengthen the muscles, so the contractions are less painful. Distraction of the brain through breathing exercises, imagining colours (especially orange), and other soothing environmental stimuli also seem to help. Giving birth to a child in a natural squatting or upright position often makes it a bit easier due to gravity. Subsequent children usually produce a shorter period of labour and less overall pain than the first birth. And if the baby is in a difficult or unusual position in utero, there are medical processes to assist with the pain. Among these are epidural blocks (spinal anaesthesia) or even Caesarean-section surgical delivery under epidural blocks or general anaesthesia.

IS IT TRUE THAT BEING MALE IS HAZARDOUS TO YOUR HEALTH?

(Asked by Sally Porter of Little Rock, Arkansas, USA)

As strange as it may seem, you could say being male is also hazardous. Yet it is not only true from birth onwards, but also from conception. In their mother's womb, males must struggle merely to become male. When they do, males can look forward to having more genetic diseases, twice the susceptibility to other diseases throughout life, higher death rates at every age of life, a nearly 10 per cent shorter lifetime, and the stigma of performing below women on virtually every behavioural indicator.

Biologically, it's far easier to become female. Developing into a female is the rule, and into a male the exception — much more has to happen for an embryo to become male than for it to become female. As the famed endocrinologist Dr Alfred Hoet said, "the basic surviving human form is female and masculinity is something additional". The struggle to become male starts just after conception. According to Dr Michael L. Gustafson and Dr Patricia K. Donahoe,[10] for a male to become a male, "a cascade of complex molecular and morphological events must occur at the appropriate time and in the correct sequence". If anything goes wrong during these events, a female develops. The sex chromosomes — X in the female, Y in the male — regulate the transformation of the fertilised ovum into an embryo. During the first 6 weeks of life, the human form tends to be female. At that time, both sexes possess identical genitalia — a primitive gonad (a sexually neutral organ) that can develop into either an ovary or a testis. If the embryo is genetically female (XX), the primitive gonad develops into an ovary without the help of any feminising hormones. If the embryo is genetically male (XY), the primitive gonad develops into a testis, but this occurs only if the "testis determining factor" (TDF) exists on the Y-chromosome. This starts the production of the male hormone, testosterone, which makes the embryo male. Testosterone enlarges the phallus, extends the urethra along its length, and closes the skin over the urogenital sinus to form the scrotum, into which the testes later descend. In the in utero struggle for gender, the basic model is female, the male an optional extra.

According to a famed anthropologist, the late Dr Ashley Montagu, it's a sad prospect to be saddled with a Y-chromosome instead of an X-chromosome. The Y-chromosome is only ⅕ the size of the X and is responsible for many genetic diseases: "That the male is endowed with a Y-chromosome seems to put him at a greater disadvantage than if he had no Y-chromosome at all."[11] There are more than 30 genetic diseases and conditions that occur more often in males than in females. They range from absence of front teeth to Van den Bosch syndrome (a condition involving mental retardation, skeletal deformities, absence of sweat glands and more). Apart from these genetic diseases, there are at least 63 other diseases to which males are more susceptible than females. But there are only 31 diseases to which females are more susceptible than males. A 1999 study surveyed all children born in Finland in 1987. Researchers found that boys have "a 20 per cent higher risk for a low 5-min Apgar score and an 11 per cent higher risk for being pre-term. After the perinatal period, boys had a 64 per cent higher cumulative incidence of asthma, a 43 per cent higher cumulative incidence of intellectual disability, a 22 per cent higher incidence of mortality and a higher, but not statistically significant, incidence of epilepsy and vision disorders." Moreover, "boys had a two- to three-fold risk of having delayed development, postponed school start or attendance in special education programmes".[12]

In Australia, the death rate for males before birth is higher than for females. Between 120 and 150 males are conceived for every 100 females conceived, but only 105 males are born for every 100 females born. We don't fully understand the reasons why males are more likely to perish in the womb. This trend continues at every year of age throughout life: in the first year after birth, for every girl baby who dies, three boy babies die; at age 21, for every young woman who dies, nearly two young men die. In Australia, life expectancy for women is 7 years longer than for men (83 years for females, 77 years for males). The 10 per cent plus difference in lifespan is nature's tax on males.

From birth onwards, males perform below females in virtually all aspects of behaviour. The exceptions are those areas requiring sheer physical strength and good muscle coordination and spatial/distance perception. For example, the hand grip of a 5-year-old boy is often twice as strong as that of·a 5-year-old girl. In the caveman era, brute strength was the go. But in our modern world, word power beats muscle power hands down every time. Females are ahead of males due to their huge advantage in verbal performance. On average boys:

☞ begin to talk later

☞ have a smaller vocabulary at preschool age

☞ begin to use sentences later

☞ learn to read later

☞ make slower progress in reading

☞ have more reading difficulties

☞ are behind girls in speed of reading, tests of opposites, analogies, sentence completion and story completion

☞ do worse in code-learning tests

☞ show a highly significant inferiority in handling linguistic relations, as in tests requiring them to construct an artificial language

☞ are less attentive to sounds

☞ learn foreign languages more slowly

☞ perform below girls in most tests of memory

☞ perform below girls in tests of logic

☞ have less vivid mental imagery.

Studies reveal that males are also more emotionally insecure. For example, if a 1-year-old male is startled by a loud noise, he tends to freeze for a short period then starts to cry. A 1-year-old female tends to be unaffected by the noise. Psychologists conclude from this that male babies seem to need more control over their surroundings. According to Dr M. Robbins, "There is broad consensus among investigators . . . that the modal female personality tends to be more socio-centric, and the modal male personality tends to be more self-centric."[13] Females are more secure emotionally and thus can afford to be social and reach out to others. Males, on the other hand, are still struggling with themselves. It's just one of many struggles in the hazardous world of being male.[14]

WHAT DETERMINES THE SEX OF A BABY? IS IT SET AT THE MOMENT WHEN THE EGG IS FERTILISED OR AFTERWARDS?

(Asked by Hugh York of Bega, New South Wales)

The sex of a baby is determined as soon as the human egg is fertilised by the human sperm. According to Dr Robert G. Brzyski from the Health Science Center at the

University of Texas in San Antonio, "Practically speaking, there is no way to modify the sex of human offspring once fertilisation occurs."[15] In humans, sex is normally determined by the sperm, not the egg. The sex chromosomes in humans are called the X- and Y-chromosomes. All normal human eggs contain an X-chromosome. Sperm can contain either an X-chromosome or a Y-chromosome. If the sperm contains a Y-chromosome, the child will be a male. If the sperm contains an X-chromosome, the child will be female. It's as simple as that. But of course, life is never that simple. In certain rare instances, the Y-chromosome in the sperm will be missing some of the genes critical to the development of the testis — the male gonad that will eventually produce sperm and the male sex hormone testosterone. In those cases the child will have female genitals, even though a routine chromosome analysis would indicate that a Y-chromosome is present. More sophisticated tests would reveal the problem in the Y-chromosome. For a woman to have functional ovaries — the female gonad that makes eggs and female sex hormones — she must have two X-chromosomes. One comes from her mother via the egg and one comes from her father via the sperm.

One's genotype is one's genetic composition. Genotype is from the Greek *genos* meaning "offspring" and *typos* meaning "type". One's phenotype is what one looks like given one's genotype. Phenotype is from the Greek *phainein* meaning "to show" and, again, *typos* meaning "type". Women who have an abnormality in one of their X-chromosomes, or who carry a defective Y-chromosome, do not have functional gonads, although their phenotype is still female. Because such women cannot make eggs they cannot reproduce in the normal fashion. And although they are considered women due to the fact that they lack male sex organs, they cannot make female hormones either. Furthermore, they do not go through the changes of puberty unless they receive hormonal therapy.

DOES THE PENIS DROP OFF IN GIRLS BEFORE BIRTH?

(Asked by Rodney Downes of New York City, New York, USA)

It is a myth that the penis develops in all embryos and then "drops off" some time before birth in females. Where did this myth come from? An important fact of embryonic development is that males and females have exactly the same body parts — including their genitals. During in utero development, some parts get big, others stay small, some fuse together, others stay separate. It is the pattern of enlargement and fusion that creates the sex differences. All embryos start with a genital tubercle. In pictures of a sufficiently immature foetus this tubercle may look like a penis. Some specialists identify this structure as the foetal "phallus". In other contexts, the phallus is just a fancy word for penis. This may be why some people get the erroneous impression that all embryos start with a penis. In biology, the phallus can refer to the penis, but it can also refer to an enlarged clitoris. This is the normal condition in some other animals. The development of the genital tubercle is why ultrasound imaging cannot be used to determine the sex of a foetus before a certain stage of in utero development. According to the definitive reference book on the subject,[16] at this early stage, both males and females look alike. In males, the genital tubercle fuses with the urethral swellings then enlarges to become the penis. In females, the structures stay separate and do not enlarge as much to become the clitoris and the labia minora. So, they're still all the same parts, just formed in a different pattern. What makes the genital tubercle fuse and enlarge in males? That's merely a question of the balance of male and female hormones. More male hormones and you get a male. More female hormones and you get a female.

ARE SOME PEOPLE BORN WITHOUT AN ANUS?

(Asked by Damien Fowler of Cessnock, New South Wales)

Surprisingly, it *is* possible for people to be born without an anus. It's more common than you would think, too. Believe it or not, about 1 in 5000 people is born without an anus. There are slightly more males than females born with this condition. There are different grades of severity and it is sometimes associated with more complex physical problems. Where necessary, it is corrected by surgery.

The anus is at the end of your intestines. More technically, the anus is the distal or terminal orifice of the alimentary canal. If an embryo does not develop intestines, it will not survive long enough to be born. Not having an anus is a somewhat different matter. An imperforate anus is a condition in which the opening of the intestines is incompletely developed, or obstructed. An imperforate anus happens early in development — only 7 weeks after conception. Four weeks after conception, the gut starts to form in the human embryo. Part of the gut is called the hindgut; in the last section of the hindgut is a cavity called the cloaca. The cloaca eventually develops into the rectum by merging with another part of the lower intestine called the urorectal septum. This is a partition dividing the cloaca into the urogenital orifice and the anal orifice. Most cases of imperforate anus are caused when the urorectal septum fails to connect properly with the cloaca. The proper partition isn't formed and the correct divisions aren't made. The bottom bit of the gut (the rectum) ends too early — somewhere inside the body cavity slightly above the anus — instead of where it should end. It is incomplete and causes varying degrees of problems based upon the degree of malformation. The result is that things don't go where they should. It's rather like a plumbing contractor who runs out of pipes when building a house — or having one that's too short.

It's the end of your end that doesn't end well.

IS IT POSSIBLE FOR TWINS TO HAVE DIFFERENT FATHERS?

Yes. There have been about 10 cases of twins born to different fathers reported in the medical literature. A recent case is that of a Croatian woman who gave birth to twins in Zagreb in June 2002. The 23-year-old woman had sex with two men around the same time. Parentage was established by DNA tests.[17]

HOW FAR APART IN TIME CAN TWINS BE BORN?

Usually it is only a few minutes. But Maricica Tescu, a 33-year-old woman from Cirlig, Romania, who has a double uterus, had twin boys delivered by Caesarean section 2 months apart! She gave birth to Catalin on 11 December 2004 and then to Valentin on 7 February 2005. Catalin was 2 months premature and weighed 1600 grams (3.53 pounds) at birth. Valentin was born at full term and weighed 2000 grams (4.41 pounds). By then, Catalin weighed the same.

WHAT'S THE WORLD'S TINIEST BABY?

The tiniest baby ever to survive is Rumaisa Rahman. She was just 244 grams (8 ounces) when born in September 2004 at the Loyola University Medical Center in Chicago. Rumaisa was 14 weeks premature when delivered by Caesarean section along with Hiba, her twin sister. Hiba was twice the weight of Rumaisa, at 563 grams (1 pound, 4 ounces), but still very small as babies go. Nine out of 10 babies weigh between 2400 grams (5 pounds, 4.6 ounces) and 4800 grams (10 pounds, 9.3 ounces) at birth. Both sisters were born with eye problems that had to be corrected with laser surgery. Such eye problems are common in premature babies. Only 62 babies weighing less than 369 grams (13 ounces) at birth have been known to survive.[18]

WHAT'S THE WORLD'S LARGEST BABY?

The largest baby born to a healthy mother was a boy weighing 10.2 kilograms (22 pounds, 8 ounces) who was born to Carmelina Fedele in Aversa, Italy, in September 1955.

WHAT IS A FALSE PREGNANCY?

False pregnancy (pseudocyesis) is a condition in which a woman is not pregnant, yet she shows physical and psychological signs of a genuine pregnancy. Pseudocyesis is also called pseudopregnancy, imaginary pregnancy, spurious pregnancy and, in less sensitive times, hysterical pregnancy. False pregnancy occurs in humans, but also in mice, rabbits, cats, dogs, goats, horses, bears and other mammals. In humans, pseudocyesis usually results from an overpowering desire to have a baby. In such cases, although a woman is definitely not pregnant, her menstrual periods cease, her abdomen gets larger, her breasts swell and may even secrete milk, her uterus and cervix may exhibit changes that happen only in a genuine pregnancy, and even urine pregnancy tests may show a false positive.

During pseudocyesis a woman may report morning sickness symptoms, food cravings, mood swings and the sensations of foetal movements. There is a general mimicking of a genuine pregnancy when pregnancy doesn't exist. A woman with pseudocyesis often remains so convinced that she is pregnant that she becomes clinically delusional. Eventually, when no baby is born, she may suffer from clinical depression. It has been theorised that such depression can alter the normal activity of the pituitary gland and cause hormone level changes that actually resemble the hormone changes of a genuine pregnancy. Dr H. Griengl reported the sad case of a patient suffering from sterility who nevertheless developed pseudocyesis. She did not have a history of any psychiatric disorder or brain damage. All she had was a strong urge for a baby that had been crushed by the sterility of her body.[19]

DO MEN EVER SHOW PREGNANCY SYMPTOMS?

They can and they often do. It's called couvade syndrome or sympathetic pregnancy. When exhibiting couvade behaviours, a father-to-be will mimic the genuine pregnancy symptoms of his pregnant partner. These symptoms include the nausea of morning sickness, food cravings and other variations in appetite, insomnia, odd dreams, mood swings, unusual emotional sensitivity, swelling of the breasts and weight gain. When his partner goes into labour, the couvade father will report labour pains too. Sometimes these imagined labour pains are claimed to be more painful to him than the real labour pains are to her. Couvade behaviours are seen in fathers throughout the world. In one study, 22.5 per cent of expectant fathers sought medical care for their "in sympathy" symptoms. Dr S. Masoni and 4 colleagues write that various couvade behaviours are seen in up to 65 per cent of men. Their study found that the couvade syndrome constitutes "a peculiar imaginary and behavioral reality" for the father-to-be. They suggest that couvade behaviours are psychosomatic in a man, "indicating an attempt to share his partner's pregnancy pain and anxiety".[20]

Canadian researchers suggest that the reason why couvade behaviours occur in some men and not in others may have to do with differences in their brains. Dr A.E. Storey and 3 colleagues write that "hormones may play a role in priming males to provide care for [their] young". They note that the hormone prolactin is important in mothering. In their experiments, the Storey team found that the brains of couvade-behaviour men are different from those of non-couvade men in that couvade men produce more prolactin when they are "exposed to a pregnancy event" — they are biochemically more prone to mothering.[21] Couvade comes from the French word *couver* meaning "to hatch". As for explaining couvade behaviours, the Canadians' Storey is the best theory hatched so far.

If you cannot stand children, you have misopaedia.

Babies are born with 300 bones. But by adulthood we have only 206 bones in our bodies.

The most foetuses to have developed in a human body were 15 — 10 girls and 5 boys — which were 4 months old (16 weeks gestational age) when they were removed from the womb of an Italian housewife in July 1971. The woman had been taking a fertility drug.

For roughly 6 to 7 months after birth, an infant can breathe and swallow at the same time. Older children and adults cannot do this.

A baby can urinate any amount at one time: sometimes they may urinate just a few drops, or they could urinate as much as 50 millilitres or more at a time. Urine output also depends on the weight of the baby. A heavier baby will urinate more than a lighter baby. A well-hydrated baby will urinate about 6 to 8 times in 24 hours.

According to UNICEF's *The State of the World's Children 2006*, in 2004, an estimated 10.5 million children died before they reached age 5, most from preventable causes. Malnutrition accounts for 5.5 million of these. Vaccine-preventable diseases account for a further 1.4 million of these.

DOES MASSAGE BOOST A BABY'S GROWTH AND DEVELOPMENT?

Research shows that infant massage stimulates the growth and development of babies, particularly premature ones. According to Dr Tiffany Field and 2 colleagues, massaging premature babies results in greater weight gain and earlier release from hospitalisation.[22] Twelve per cent of US infants are born prematurely and 8 per cent are of low birth weight. These babies are more likely to die in infancy, suffer physical and mental problems and require expensive hospitalisation. In 1 study it was found that, on average, premature babies who received a 15-minute massage 3 times per day had a 47 per cent greater weight gain and a 4.6-day shorter hospital stay than premature babies who were not massaged. The shorter hospital stay means that not only are the massaged infants healthier, but thousands of dollars may also be saved in hospital care for each tiny patient. Dr Field believes that regular massages improve an infant's nervous system and increase production of hormones that allow them to absorb more food and hence grow better.

In another Field team study, mothers gave asthmatic children 20-minute massages every evening at bedtime for a month. It was found that the anxiety levels of both the children and the mothers decreased. The children's moods improved and their levels of the stress hormone cortisol decreased. Most importantly, over the 1-month period, the children had fewer asthma attacks and were able to breathe better based upon daily peak airflow readings.

In still another study, the Field team wanted to determine whether massage therapy improved the classroom behaviour of preschool autistic children. After 1 month of massage therapy, the autistic children were less touch-sensitive, less distracted by sounds and more attentive in class, and performed better on tests of sensory and social behaviour.

In yet another study, this time dealing with diabetic children and the parents who must treat them (often with painful insulin injections several times a day), the Field team showed that a 1-month period of massage therapy improved children's blood sugar readings "to the normal range". It was also found that immediately after the massage therapy sessions the anxiety and mood levels of both children and parents had "abated significantly".

In a final study, the Field team studied the effect of massage in reducing post-traumatic stress disorder (PTSD) on Miami school children caused by the devastating disaster of Hurricane Andrew. School children were massaged twice a week at their school for 1 month. After the month, it was discovered that the massaged children had less depression and anxiety and lower cortisol levels. In addition, their drawings had fewer depressive and disorganised features, which can indicate PTSD.[23, 24]

Some findings from other studies:

Massaged premature babies performed better on the Bayley Mental and Motor Scales, indicating superior mental and physical development than non-massaged infants.

Cocaine-exposed newborns are at a high risk of suffering a whole host of neurological problems. In a study of such newborns, massaged babies showed fewer neurological problems on the Brazelton Scale test and enjoyed a 28 per cent greater weight gain than non-massaged newborns.

HIV-exposed infants experience more developmental delays than infants not exposed to HIV. Such babies often never catch up physically, mentally and behaviourally to normal babies. In a study of these infants, massaged babies performed better on the Brazelton Scale than non-massaged infants exposed to HIV.

Thus it would seem that infant massage rubs everyone the right way.

CAN A MAN HAVE A BABY AFTER HE'S HAD A SEX-CHANGE OPERATION?

When men have a sex-change operation, it is only the genitals that have been changed. They cannot have babies because they still do not have a female reproduction system (FRS). A sex-change operation is more correctly termed sexual reassignment surgery (SRS). When the change goes from male to female it is called vaginoplasty. It is a dramatic and irrevocable final step in the male-to-female gender transition. It is taken only after the deepest introspection by the individual, who undergoes counselling regarding all the options available. For those needing a complete gender correction, many experts regard this surgery as a life-saving and life-enhancing procedure that enables an individual to live a happier life. Other experts argue that the procedure can be mistakenly seen as a panacea for individuals with complicated physical and psychological problems far beyond those of gender. They warn that an SRS can sometimes lead to permanent and terrifying emotional and psychological consequences.

Apart from this debate, in order to have a baby, a male would need to have a fully functioning FRS. Although extremely difficult, surgical transplantation of certain parts of the FRS into the male is feasible. But such a transplanted system could not function naturally even if the FRS could be successfully transplanted in its entirety. The FRS is controlled by hormones, and many of these are produced in other parts of the female body. For example, for a transplanted FRS to work naturally in a male, two parts of the female brain — one from the hypothalamus and one from the pituitary — would also have to be transplanted into the male brain. The hypothalamus and pituitary in females control ovulation in the ovaries. No ovulation, no pregnancy. No pregnancy, no baby.

HOW LOUDLY CAN A BABY CRY?

Pretty loudly! A baby can produce a cry as loud as 96 decibels. Workplace regulations on construction sites limit noise exposure to about 85 decibels. Higher than that and workers risk permanent hearing loss. Parents often hold their crying baby on their shoulder next to their ear. This is the worst place to hold them from the standpoint of the parent's ears (and nerves). Dr Mats Zackrisson of the University of Gothenburg in Sweden claims that, although it is theoretically possible that parents may risk hearing impairment from a crying baby, it is not very likely. Dr Zackrisson calculates that exposure to a baby crying at 96 decibels for 8 hours every day for 30 years would be needed in order for 40 per cent of parents to suffer "considerable hearing impairment". So a crying baby who damages your hearing is nothing to cry about.[25]

WHAT DOES IT MEAN WHEN A BABY CRIES?

(Asked by Lindy Williams of Croydon, New South Wales)

Crying is an expression of fatigue, sadness, fear, loneliness, frustration, anger, discomfort (too much heat or cold), pain, boredom or hunger. Hunger is the most common reason for newborn crying. Boredom is probably the one least anticipated by parents. If a baby is bored in their crib, for example, pictures on the wall and ceiling, mobiles etc may provide a more stimulating environment and reduce boredom. But are you leaving the baby alone in their crib too often or too long? Babies are social from birth. Most researchers believe that the cry of the newborn is only an indicator of distress. The intensity of the crying will vary depending whether the cry represents an emergent or an urgent need to be fed, or some acute distress such as pain. If the need expressed is not immediately or soon met, the cry may become more intense.[26]

DOES MY BABY CRY TOO MUCH?

(Asked by Lindy Williams of Croydon, New South Wales)

Research shows that infants spend about 6 to 7 per cent of their day crying. But there is tremendous variability in the amount of crying from infant to infant and from day to day. For instance, the infant may react to a soft toy with interest and pleasure if they are in an alert state. But the baby may react to the same toy with crying if they are tired.[26]

IS CRYING CONTAGIOUS?

(Asked by Lindy Williams of Croydon, New South Wales)

In a sense crying *is* contagious. Studies show that infants less than a day old will cry when they hear a recording of another infant crying. But interestingly, they will not cry if they hear a recording of their own crying. Some researchers have interpreted this as an indication of an inborn capacity of humans to react to the suffering of others.[26]

IS IT POSSIBLE TO PREDICT IF I WILL HAVE A BABY WHO CRIES A GREAT DEAL?

(Asked by Lindy Williams of Croydon, New South Wales)

A study by Dr Ian St James-Roberts and Dr P. Menon-Johansson attempted to predict infant crying from foetal movements. They had 240 mothers note down how often their babies kicked in utero and also observed babies in utero with ultrasound. The researchers suggest that the amount of infant crying is determined before the baby is born. They could only speculate as to why.[26, 27]

CAN PARENTS, MOTHERS IN PARTICULAR, DISTINGUISH THE CRY OF THEIR OWN INFANT FROM THE CRY OF OTHER INFANTS? WHEN DOES THIS START?

(Asked by Lindy Williams of Croydon, New South Wales)

Yes, they can. But this ability does not begin at birth. It has been reported in the child-development research literature that as early as the fourth day of life most mothers will be able to differentiate the cries of their own babies from those of other infants. Mothers become increasingly skilful at distinguishing between cries of hunger, fatigue, pain or anger.[26]

ARE THERE TYPES OF CRYING?

(Asked by Lindy Williams of Croydon, New South Wales)

Four types of crying have been distinguished by child-development specialists. These are:

- the hunger cry, which is a rhythmic cry that starts with a whimper and becomes louder and more sustained;

- the fatigue cry, which is much like the hunger cry but is a response to being tired rather than being hungry;

☞ the pain cry, which begins with a shriek, is followed by a second of silence as the baby takes a breath and then becomes more vigorous; and

☞ the anger cry, which is also rhythmic but much more intense.[26]

IF PARENTS RESPOND IMMEDIATELY TO THEIR INFANT'S CRY WON'T THAT "SPOIL" THE BABY?

(Asked by Lindy Williams of Croydon, New South Wales)

A prompt response to an infant's cry does not encourage more crying. In fact, evidence in the field of child development since 1972 has established beyond a doubt that a prompt response to an infant cry lessens infant crying and even has the added benefit of helping the baby develop communication and language skills earlier. The result of studies on this topic show that in the first year or so, children — still dependent and needy — cannot be "spoiled" by parents who are responsive and attentive.

In fact, it's quite the reverse. Babies become more secure, compliant and independent when their behaviours indicating distress are responded to without delay. Babies cannot talk, write a letter of complaint, or get out of the crib and head for the pub. All they can do is cry when they are afraid, lonely, upset, too hot, too cold, hungry, or thirsty. When babies are responded to, they grow up feeling the world is a secure, safe, responsive place. When babies do not have these basic needs met, they become insecure, complaining, demanding, and far more difficult to manage.

Some babies give up crying altogether if no one responds. This almost certainly will have disastrous consequences, as the baby is establishing the first bonds of trust and self-esteem at this age. If they find that no one responds to them, they learn that the world is a hostile place — and that they are unworthy of love.

It is a myth that a baby is sophisticated enough to manipulate a parent. All they are doing is expressing themselves as best they can to have their legitimate needs met.

SHOULD PARENTS TRY TO ANTICIPATE THE INFANT BEFORE IT CRIES?

(Asked by Lindy Williams of Croydon, New South Wales)

Intelligent parents try to stay one step ahead of their child. Part of this is anticipating the infant's needs. The parent who responds to the restlessness of the infant before they cry may be spared the more energetic protests that evolve if the infant's needs remain unmet. After all, the baby cannot speak to you. The best they can say is, "Whaaaaaa!"[26]

HOW DO YOU BEST SOOTHE A CRYING INFANT?

(Asked by Lindy Williams of Croydon, New South Wales)

Some infants are easier to soothe than others. There are several ways to soothe a crying infant. If the baby is hungry, obviously feed them. Picking up the baby and putting them on your shoulder often changes the crying state to a state of alert inactivity. Studies of parents who rock, hum or rhythmically stimulate their infant in various ways have found that such stimulation has the effect of reducing the baby's activity level and decreasing both the heart rate and the respiratory rate.[26]

DO BABIES ALL OVER THE WORLD HAVE THE SAME CRYING PATTERNS?

Research reveals that there are cross-national differences in infant crying patterns. For example, Japanese babies cry very little compared with European babies. This has been observed for many years by researchers and is probably due to the fact that Japanese mothers attend to their infants very quickly. In one famous study of the 1940s, it was found that Russian babies who were kept tightly swaddled cried very little, but also smiled very little when tickled.[26]

ARE THERE ANY ABNORMAL FORMS OF INFANT CRYING?

Although very rare, there are some abnormal forms of infant crying. If they are present from birth, it could indicate that the infant has a serious health problem. In fact, one particularly tragic form of mental retardation is named for the characteristic cat-like cry that those suffering from the syndrome make. It is called *cri du chat* syndrome, which literally means "cry of the cat".[26]

WHAT IS "MOTHERESE"?

Motherese is the name given to the language we used during babyhood and early childhood. It's found in speakers of Arabic, Chinese, English, French, Japanese, Russian, Spanish and probably every other language too. Motherese explains why "y" is added to most everything when we are talking to babies. "Dog" becomes "doggy", "pig" becomes "piggy" and so on. Major features of motherese include the exaggerated rise in the speaker's pitch and intonation when speaking to a baby, simpler and more concrete vocabulary, shorter and more grammatically correct sentences, and the avoidance of past tenses and pronouns. Motherese uses more questions, more commands and more repetition. We use fewer disfluencies too. These are sentence interrupters such as "um" and "ah". In motherese we're also more likely to choose onomatopoeic words such as "choo-choo train" instead of just "train".

CAN A BABY CRY IN THE WOMB?

(Asked by Lindy Williams of Croydon, New South Wales)

An unborn baby can be seen to apparently "cry" at as early as 21 weeks gestational age. They seem to make crying movements, but do not make any crying sounds. Certainly by the third trimester of pregnancy (26 to 40 weeks gestational age), the baby has vocal cords capable of making sounds. But a baby in the womb is totally immersed in a salty solution that it continually swallows and "breathes", called amniotic fluid. While in the womb, there is no air with which to make crying sounds. Only after a baby is born can its lungs become fully expanded to take in air. To create sounds with our vocal cords, air is necessary to move the sound from our lungs through the windpipe, past the vocal cords and out the mouth. Even when we make vocal sounds underwater while swimming, the sound is made by pushing out air in our lungs through our vocal cords. Therefore, although an unborn baby is physically capable of crying, it doesn't cry in the full sense of the word because no sound can be heard.

WHAT IS "CONTROLLED CRYING" AND SHOULD I USE IT ON MY BABY WHO HAS TROUBLE SLEEPING?

(Asked by E. Kocabas of Brooklyn, New York, USA)

This would normally be a question for a parental advice columnist. But since the answer involves the developing human brain, here goes. Between 20 and 45 per cent of parents report that they have problems getting their newborn baby to sleep through the night. The baby often wails away at, say, 3 am, making the weary parents even wearier. This is because the baby, newly out of the womb — that warm, dark, wet little space with room service 24/7 — has not become used to the adult human's normal pattern of sleeping through the night. The brain of a newborn is incapable of immediately adopting the adult sleeping patterns.

"Controlled crying" is a technique invented about 20 years ago by a Boston paediatrician to get babies to sleep through the night. Once the baby is put down to sleep, it involves letting the baby "cry it out" without parental attention for progressively longer intervals of time until the baby does not cry at all. Many parents say that "controlled crying" succeeds in getting their baby to sleep through the night. However, many parents say "controlled crying" doesn't work for their baby and sometimes it even makes matters worse. The parents themselves sometimes "stress out" too, as they are forced to seemingly go against nature and *not* tend to their baby's obvious upset.

Some experts warn that allowing a baby to "cry it out" causes extreme distress to the baby. Extreme distress in a newborn has been found to block the full development of certain areas of the brain and cause the brain to produce extra amounts of cortisol, which can be harmful. According to 1 study, children who suffer early trauma generally develop smaller brains.[28] Another study published in the same journal claims that the brain areas affected by severe distress are the limbic system, the left hemisphere and the corpus callosum. Additional areas that may be involved are the hippocampus and the orbitofrontal cortex.[29] Thus, with "controlled crying" you may get the baby to sleep, but at what cost? Alternative "gentle baby sleep" techniques are available on the Web.[30]

IS SKIN-TO-SKIN CONTACT GOOD FOR MY NEWBORN BABY?

(Asked by E. Kocabas of Brooklyn, New York, USA)

This is also more a question for the parent advice columnist. But it also involves the body. (It is also a slam dunk to answer, as all experts agree.) Early skin-to-skin contact involves placing the naked baby on the mother's bare chest at birth or soon afterwards for at least 24 hours. It is greatly beneficial for both the baby and the mother. This was demonstrated recently in an analysis of 17 studies on the topic. G.C. Anderson and 3 colleagues write that "This could represent a 'sensitive period' for priming mothers and infants to develop a synchronous, reciprocal, interaction pattern, provided they are together and in intimate contact. Routine separation shortly after hospital birth is a uniquely Western cultural phenomenon that may be associated with harmful effects including discouragement of successful breastfeeding." The researchers found that the studies suggest early skin-to-skin contact has 2 principal benefits: it facilitates successful breastfeeding and it helps reduce infant crying. Furthermore, they write, skin-to-skin contact "has no apparent short or long-term negative effects". Bet it feels good too![31, 32]

WHEN CAN THE FOETUS FEEL PAIN?

The foetus can feel pain at no earlier than 20 weeks gestational age. If anything, this is a conservatively late estimate. Components of the spinothalamic system, which signals pain throughout the body, begin to be built at 7 weeks gestational age. By 12 to 14 weeks gestational age, sufficient development has occurred to make some pain perception likely. By 20 weeks gestational age the spinothalamic system is fully established and connected. According to Dr Paul Ranalli, a neurologist at the University of Toronto, unborn babies between 20 and 30 weeks gestational age may actually feel pain *more* intensely than full-term babies after birth, children or adults. This is because the spinothalamic system is fully developed, yet the body's natural analgesic system, which fights pain, is somewhat less developed.

CAN NEWBORN BABIES ANTICIPATE PAIN?

It seems they can. Newborn babies can learn to anticipate pain and suffer more intensely as a result. This is according to a Canadian study that looked at pain in full-term babies who were born to mothers with diabetes. These babies had to undergo repeated heel lances as part of the testing of their blood glucose level in the first 36 hours of life. Their pain from the procedure was compared with that of another group of newborns matched in terms of birth weight, sex, method of delivery and anaesthesia administered during labour. Babies in the diabetic mothers group had to have their skin cleaned prior to each blood test. They subsequently showed that they anticipated pain as soon as they felt the skin-cleansing solution being applied. Videotapes of the babies revealed that the diabetic mothers group showed higher pain scores based on their crying and grimacing. According to the researchers, "After an average of 10 heel lances (preceded by skin cleansing), infants responded to cleansing of the skin with behaviors indicative of pain."[33, 34]

DOES BIRTH TRAUMA AFFECT US THROUGHOUT LIFE?

(Asked by M. Downes of White Plains, New York, USA)

There is plenty of evidence that when our birth doesn't go well it can affect us later in life — and that effect is not very pleasant. Birth trauma can lead to teenage suicide, drug addiction and a life of violent crime.

☞ A 1985 study found that 1 of 3 main predisposing factors for teenage suicide was "respiratory distress for more than one hour at birth". The other 2 were "no antenatal care before 20 weeks of pregnancy" and "chronic disease of the mother during pregnancy".[35]

☞ A 1990 study found that if mothers were given barbiturates, opiates or large doses of nitrous oxide during labour, the risk of their child's becoming addicted to opiates later in life was increased 4.7 times.[36]

☞ In 1997, a study of 4269 consecutive births at a single hospital found that when combined with the mother's extended separations from her baby or the rejection of her baby altogether in the first year of life, birth complications "predispose an individual towards violent crime". One factor alone does not affect the risk of committing violent crime, but the two together were "almost like a chemical reaction" for violent crime.[37]

☞ Dr Michel Odent writes that when he visits a foreign city, in order to judge whether or not it is safe for him to walk the streets at night, he looks at the local birth statistics. He says: "My rule of thumb is that the rates of criminality are correlated with the rates of obstetrical intervention. This means, for example, that I'll be extremely cautious in places such as Sao Paulo, Mexico City, Rome or Athens, where the rates of caesarean sections are astronomical. On the other hand I'll be more relaxed on the streets of Tokyo, Stockholm or Amsterdam, where they have maintained comparatively low rates of obstetrical intervention."[38]

ARE LARGE BREASTS NEEDED TO FEED A BABY?

Adequate breasts are required, but not particularly ginormous ones. Breast-producing (adipose) tissue is not required in large supply for normal breast-feeding (lactation). In fact, women nursing their babies for 6 months or more can undergo a reduction of breast volume compared with their pre-pregnancy size and still produce an adequate amount of milk. The greater amount of adipose tissue in women than in men is undoubtedly related to the differences in sex hormones. Adipose tissue increases in volume in response to the female hormone oestrogen. An unsolved question is why fat cells in the breast are so much more responsive to oestrogen than other cells quite close by in the skin.[39]

WHY AREN'T GIRLS BORN WITH LARGE BREASTS?

They don't need large breasts until they become mothers. Before puberty, they'd just get in the way.

DOES INCEST ALWAYS PRODUCE CHILDREN WITH GENETIC DEFECTS?

(Asked by S. Donohue of Hamilton, Ontario, Canada)

The prohibition of incest is found in all human societies. However, different societies define incest differently. All societies ban marriages between parents and their children and between siblings. Beyond this, societies differ as to what is allowed, particularly as to cousin–cousin marriage. There are also many instances where the incest taboo has been ignored. In ancient Egypt, Cleopatra was the product of 11 generations of royal marriages between brothers and sisters. She attributed both her beauty and her intelligence to her so-called "pure bred" royal lineage. The European royal families continued to marry mostly within their own extended families until well into the last century.

There are at least 4 theories as to why restrictions prohibiting marriage within a family group (endogamy) have arisen. The Genetic Defect Theory holds that incest can result in genetically defective offspring due to the lack of variability in the gene pool. An example of this is the genetic-based disease haemophilia that has afflicted the European royal families. It has even been suggested that this was an indirect cause of the Russian Revolution in 1917. A genetic disease is more likely to show itself when endogamy is practised rather than when exogamy (marrying outside the family group) is practised. However, good genetic characteristics can come about through endogamy as well as bad. This forms the basis of racehorse, dog and cat breeding, as well as livestock breeding for farming purposes.

Most experts today prefer other explanations for the incest taboo. The Social Role Theory holds that marriage and kinship systems are important for the assignment of clearly defined social roles. When closely related kin marry, these roles become confused and society breaks down. The Jealousy Theory holds that marriage between close kin encourages family jealousy, suspicion, rivalries and family disharmony. The Alliance Theory holds that marriage inside the immediate kin group fails to achieve vital cultural, economic, political and warfare/defence relationships with neighbouring groups. An incest taboo forces people to go outside their own kin group for marriage partners, which helps improve relations with neighbours.

Whatever theory you subscribe to, the incest taboo makes people, families and society stronger.

WHY DO WE HAVE 2 OF SO MANY BODY PARTS?

(Asked by Richard McLain of Cardiff, New South Wales)

Our body is made up of many sets of 2 body parts: we have 2 eyes, 2 ears, 2 hands, 2 feet and so on. Often, as the saying goes, 2 are better than one. We see better with 2 eyes because our depth perception is improved through triangulation (using the angle created between an object and 2 fixed points to perceive distance). We hear better with 2 ears for the same reason. When there is only 1 body part, the mouth or the nose, for instance, usually it is located in the middle of the body, where it will do the most good and also not be far away from any other part of the body.

If you split your body in half right down the middle, each half would look nearly identical to the other half. This is called bilateral symmetry. But each half actually differs slightly. You can see this for yourself if you look closely at the 2 sides of your face, your 2 hands and your 2 feet. You'll notice each is a little different. If you photograph half of your face, then reverse the image, then put both halves together, you will probably not recognise the face in the photo as you.[40]

IS THERE ANY PART OF THE BODY WE DON'T NEED?

(Asked by Jake Donner of El Paso, Texas, USA)

Various parts of the body have been nominated as obsolete and useless (vestigial). But we might think twice before putting any body part on human evolution's delete list. One vestigial candidate is the appendix. We can live without it, yet it has a function in the human immune system. It is located at the entrance of the almost sterile ileum near the colon, where it helps lower bacteria count. What about tonsils? We can live without them too, but they also help the immune system by blocking bacteria from reaching the entrance to the cavity at the back of the mouth (pharynx) and the oesophagus. The tiny pineal gland located in the brain was once thought useless. But we now recognise that it produces two important hormones, melatonin (which helps keep normal body rhythms going) and serotonin (which helps keep brain neurons functioning normally). Although it gets smaller as we mature and might at first seem useless, the thymus is a small organ near the neck and heart that is important to the lymph system and in which vital T-lymphocytes multiply and mature in order to fight disease. Wisdom teeth are used for grinding food. Given our modern diet, we are less likely to eat raw meat, have dirt and sand in our vegetables and so on. Lots of grinding is less important and wisdom teeth are less essential. But according to the American Dental Association, "Wisdom teeth are a valuable asset to the mouth when they are healthy and properly positioned."[41]

The best candidate for a vestigial body part might be the palmaris longus. This combined muscle and tendon in the forearms is not even present in about 13 per cent of us. Some of us have it, but only in one forearm, so living without it is already being done. It was possessed by all of our non-human primate ancestors. It strengthens the grip and thus may have been important when we needed to swing from trees à la Tarzan. But other than Olympic athletes and circus performers, few of us swing (at least not in that way). One advantage of having this muscle and tendon combination is that the median nerve is better protected from an injury to the wrist. If ever we need a tendon graft, the palmaris longus often gets the call. Thanks to technology, though it's lost its old job, it has a new one. [42-44]

HOW COULD YOU IMPROVE THE HUMAN BODY?

We humans would look considerably different — both inside and out — if evolution had designed the human body to work better and last longer. A recent study speculated that a human "better designed for old age" might possess the following features:

☞ Shorter stature would provide a lower centre of gravity, which could help to prevent the falls that often plague the elderly.

☞ More ribs could help prevent hernias and other problems of holding organs in place more effectively.

☞ Thicker bones would protect against breakage during falls.

☞ Knees able to bend backwards would make the bones less likely to grind and deteriorate.

☞ A forward-tilting upper torso would relieve pressure on the vertebrae, thus lowering the risk of ruptured disks.

☞ A curved neck with enlarged vertebrae would counterbalance the tilted torso and enable the head to stay up and forward.

☞ Thicker disks would resist pressure on the back.

☞ Extra muscles and fat would add weight to the bones, which would help counter the effects of osteoporosis (demineralisation of bones).

☞ Leg veins with more "check valves" would combat the development of varicose veins.

☞ Larger hamstrings and tendons would help support the leg and hip, thus reducing the need for knee and hip replacement surgery.

☞ If the optic nerve were attached to the back of the retina it might stabilise the retina's connection to the back of the eye, thus helping to prevent retinal detachment.

☞ A raised trachea would help food and drink bypass the windpipe and reduce choking.

☞ An enlarged mobile outer ear would collect sound with greater efficiency, to compensate for internal ear breakdowns.

☞ More plentiful and durable hair cells in the ear would help preserve hearing late in life.

With all of these changes our bodies would be better constructed, but wouldn't win any beauty contests.[45]

WHAT HAPPENED TO ALL THE OTHER HUMAN-LIKE SPECIES SUCH AS THE NEANDERTHALS? WHY DIDN'T ANY OF THESE HUMAN OFFSHOOTS SURVIVE?

(Asked by Peter Thomas of Collingwood, Victoria)

Anthropologists claim that the fossil record tells us multiple species of human-like hominids commonly shared the Earth's environment over most of the past 4 million years. According to Dr Ian Tattersall, Curator of Anthropology at the American Museum of Natural History in New York, "being alone in the world as a species, as we are today, is the exception, not the rule. The only obvious reason for the current state of affairs is that the human is an unusual creature — intolerant of competition by close relatives and able to do something about it."

The fossil record shows that between 2 and 3 million years ago, at least 4 species of hominids coexisted on the shores of what is now Lake Turkana in Kenya. Hominids are human-like creatures that walk upright on 2 legs and comprise both the humans of today and extinct ancestral human-like creatures. Hominids are part of a larger category, hominoids, which also includes non-human-like primates such as pongids (modern chimpanzees, gorillas and orang-utans), gibbons and extinct ancestral and related forms such as Proconsul and Dryopithecus. Proconsul africanus lived as late as 14 million years ago and may have been the ancestor of modern chimpanzees. Dryopithecus was a creature with a semi-erect posture lived as late as 20 million years ago and could be an ancestor to both man and modern apes other than chimps.

Hominids ventured from Africa, perhaps from elsewhere too, and diversified wherever they spread. Those similar to today's humans emerged between 100,000 and 200,000 years ago. They reached Europe about 40,000 years ago and Australia at about the same time. They brought with them sculpture, painting, religion, simple notation, music and a host of other aspects of culture. However, when they arrived, Europe was already occupied by Neanderthals. Tragically, within 10,000 years the Neanderthals were gone — almost certainly as a result of failing in competition with the newcomers. Recent dating of fossils in Java suggests something similar may have happened in east Asia.

HOW WERE NEANDERTHALS DIFFERENT FROM US?

(Asked by Luke Stewart of Manorhaven, New York, USA)

It seems questions about the Neanderthals simply will not go away. People are very curious about this extraordinary human ancestor. In 1856, at the Feldhofer Cave in the Neander Valley (*Neander Thal* in German) near Dusseldorf, Germany, workers discovered the bones of an early form of modern human who had lived no later than between 30,000 and 60,000 years ago. Compared with today's humans, this early human was shorter, squatter, larger, heavier, had a bell-like chest cavity, a wider pelvis, was very compact, and effectively had no waist. This body shape helped Neanderthals adapt to the very cold climate in which they lived. Ice was perhaps 305 metres (1000 feet) thick all year round on the top of the plateaus that surrounded the valley in which they lived. They had flatter noses, larger nostrils and larger nasal cavities than our own, which allowed for better warming of air as they breathed it in. The skulls of the Neanderthals had lower, more sloping foreheads than ours. Their jaws were heavier and also stronger; they could probably bite through a chicken leg. Their brains were longer, positioned lower in the skull, and rested more behind as well as more above their faces. It is a myth that the

Neanderthals had smaller brains than ours — in fact, their brains were of equal size or even somewhat larger. Although their brains would probably have supported speech, their nasal passages would not, so Neanderthals probably did not speak as clearly as we do. They probably grunted. Their bones were much larger and much thicker than ours. Their hands, legs and feet were particularly sturdy. They were extremely muscular and had the skeletal frames to support their great muscle weight. Mr Neanderthal would be Mr Universe today.

Neanderthals lived as hunters and gatherers. Life expectancy was about 40 to 45 years. They lived in groups of 30 to 50. They had stone tools, but little technology beyond this. Neanderthals cared for their sick and injured. They buried their dead with ritual and therefore must have had some form of religion. Neanderthals built homes out of wood, bone and animal skins. Some homes were 10 metres (30 feet) long and 3 metres (10 feet) high. Neanderthals have left only a few items of artwork on bones and rocks.

Many theories have been put forward as to why Neanderthals disappeared. They may have simply bred more slowly than their successors (Cro-Magnon Man); died out simply because of their shorter life expectancy; been wiped out due to introduced diseases (such as what happened to native North and South Americans after first contact with Europeans); or been exterminated by their successors in an early form of ethnic cleansing. We don't know.[46]

WHEN EARLY HUMANS EVOLVED, WERE THEY HUNTERS FIRST THEN GATHERERS OR GATHERERS FIRST THEN HUNTERS?

(Asked by T. Jordan of Gainesville, Florida, USA)

Hunting and gathering probably emerged together at the same time. But before either hunting or gathering emerged, early humans were probably scavengers. A 2005 book argues that there was even a step before scavenging: humans were prey. They were high up on the menu for many predators. This is the theory advocated by Dr Donna Hart and Dr Robert Sussman.[47]

Most people today probably think that humans have always been at the top of the food chain. With their superior intelligence, they were supposed to have ruled above all other animals. Not so, say Dr Hart and Dr Sussman. They present impressive fossil evidence that for all but the last few thousand of the 3 million years of human existence, humans were continually trying to avoid being eaten by bigger, stronger, swifter and fiercer animals than themselves. Dr Hart and Dr Sussman relate how, for most of human existence, humans were choice menu items for a wide range of animals that included mammals such as bears, lions and tigers; reptiles and amphibians such as saltwater crocodiles; fish such as sharks; and even giant birds that are now extinct. Compared with these predators, humans were weak-muscled, slow-running, non-fanged and talonless animals who were not very effective competitors for food and rather poor in defence against predation.[48] About 100,000 years ago, humans developed our very large brain. With it, so the argument goes, humans gained dominion over all other creatures through applying that superior intellect by learning how to band together in groups, hunt and forage cooperatively, make weapons and basically out-think the predators who were feeding on them. Before these developments, the family that "preyed" together did not stay together — it was more likely to be another animal's lunch.

WHAT IS "SINGULARITY" AND WILL IT CHANGE WHAT WE ARE AS HUMAN BEINGS?

Singularity is the scientific term that refers to the point in the future when technological change will be so rapid that it will change *everything*. At that time things and events will move so fast that we will not possibly be able to predict what's coming next. We won't be able to predict short- or long-term results and certainly not ultimate consequences. According to Dr Ray Kurzweil,[49] singularity will be a time when technology and nature become one and the very definitions of life, nature and human become altered forever. It will be a time when technological change is so "rapid and profound it could create a rupture in the very fabric of human history. Singularity is the opening of Pandora's Box."

When will this era of singularity occur? According to Dr Kurzweil, some time within the next 25 years. Some of the predicted features of singularity that will directly affect the human brain include:

Artificial brains will merge with human brains, just as two computers can now upload and download information to each other. All humans will have immediate and convenient access to all knowledge of all subjects. Such an innovation will revolutionise education, schooling, and the role of teachers.

Machine intelligence will exceed human intelligence. Machines with such intelligence will deduce, create and even feel. They will demand legal rights, seek recognition and even claim to possess souls.

Nanocomputers (computers operating on a microscopic scale) will be placed in our brains via injection to enhance thinking, memory, visualisation, and emotion. They will also stimulate the brain's pleasure and pain centres. This will enhance human sensation and enjoyment while eliminating all forms of pain.[50, 51]

Aboriginal people have lived in Australia from 40,000 years ago. As such, they could be judged to be the most ancient people still in existence.

Ꮭ Ꮩ

The first known contraceptive was crocodile dung, used by the Egyptians in 2000 BC.

Ꮭ Ꮩ

If the entire population of China walked past you in a single file at the rate of 1 per second, the line would never end because of the nation's rate of reproduction.

∼ CHAPTER 2 ∼
THE HEAD

So many marvellous things have been said about the head. One of Benjamin Franklin's many pearls is "If a man empties his purse into his head, no one can take it from him." An interesting view of knowledge certainly — one that sticks in your head. When one considers the head, one naturally considers the face too. "I have always considered my face a convenience rather than an ornament," said Dr Oliver Wendell Holmes. Or as W.H. Auden described himself, "My face looks like a wedding-cake left out in the rain." Most of us do not have the countenance of a Hollywood movie star. It is highly unlikely that anyone will ever say of most of us, as Christopher Marlowe wrote of Helen of Troy: "Was this the face that launched a thousand ships?" Still, our face is the most distinctive and most recognisable feature about us. We should glory in that and be happy with our uniqueness. And now let's move a-head.

HOW MUCH DOES MY HEAD WEIGH?

(Asked by Nathan Schuster of West Richmond, South Australia)

This is not an easy question to answer. Experts can't agree on how to take the measurements. Heads also come in a variety of shapes and sizes. Of course, you could always find out your head's weight by cutting it off and placing it on a scale. Or you could slowly lower yourself onto a scale while hanging upside down from your ankles. But you might not want to do either of these.

Some experts claim you can get a good estimate of the weight of your head by measuring its volume. Assuming that your head is a sphere, it would hold about 3 litres (6.4 pints). Water weighs 1 gram per millilitre. Bone and fat weigh less than water. Therefore, your head weighs about 3 kilograms (6.6 pounds). Of these 3 kilograms, the brain weighs about 1.5 kilograms, the skull about 1 kilogram and the skin about 500 grams. Disputing this, the editors of *New Scientist* magazine attempted to calculate the average human head as weighing 4.25 kilograms (9.36 pounds).[1] Still others calculate 4.5 to 5.4 kilograms. It could be a little more or a little less depending upon your hat size. Oh yes, and if you count your ears, that's extra.

IS THERE A DISEASE THAT SHRINKS THE HEAD?

It is sometimes said of Wegener's granulomatosis that it "shrinks" heads and also "dissolves" noses, ears and other parts of the body. Wegener's granulomatosis is a strange disease of unknown origin that starts in the respiratory tract and moves up the body to the neck and head. It inflames arteries and blood vessels and cuts off the blood supply to essential body organs. Organs then dissolve, leaving big cavities that can look truly gruesome. People can lose their noses or ears and their heads can appear to collapse. Two Italian rheumatologists, Dr V. Riccieri and Dr G. Valesini,[2] report that in the past a person with Wegener's granulomatosis usually died after suffering from the disease for about 5 months. But now, with modern drugs, the disease can be effectively treated and the patient cured.[3]

IS IT TRUE THAT HUMANS ARE THE ONLY ANIMALS THAT CAN LEARN TO STAND ON THEIR HEADS?

Almost. Don't forget elephants.

CAN I TELL HOW MY FACE WILL AGE?

Time, sun and gravity take their toll on our faces. Tissue sags, wrinkles, and loses its youthful tone. Facial lines (rhytids) can be shallow or deep. Facelift surgery is called a rhytidectomy. We now have the technology to fight Father Time.

How will your face age? Dr Michael Bermant, a plastic surgeon in Chester, Virginia, suggests you sit down with a mirror and a photograph of yourself as a young adult and look at both closely. Then look at photos of your parents and older family members when they were your age and compare them with how they look now. This will give you some idea of how your face might age.

What normally happens in an ageing face includes:

☞ The ageing forehead and brow can show sagging of tissue and lowering of the eyebrows. Horizontal furrows and folds can appear. Vertical lines between the eyebrows can appear too. There can also be an elevated hair line, the appearance of loss of hair and the actual loss of hair.

☞ The ageing eyelid tissue can appear to have some folding of unnecessary tissue of the upper eyelid, making it difficult to keep the eyelid open. There can be a weakening of support for the lower eyelid, which can let the lid sag away from its protective position against the eye, sagging of skin and muscles of the lower lid, and tissue folds (festoons) above the cheek.

The transparency of the eyelid skin can reveal the darkness of the eye socket behind it. There can be wrinkling of skin at the outer corners of the eyes extending towards the hairline — the dreaded "crow's feet".

☞ The ageing cheek suffers a downward migration of facial structures from the cheekbones, resulting in less prominent cheeks. There can be a deepening of the lines between the nose and the corners of the mouth (nasolabial folds). There can also be tissue below the jaw line, creating bulges (jowls).

☞ The ageing neck has unnecessary tissue. When excessive, this produces a turkey gobbler-like wattle from sagging neck muscles, which can tighten into bands. These bands can blunt expression of the angle between the chin and the neck, resulting in a loss of jaw definition.

DO FACIAL FEATURES AFFECT PERSONALITY?

(Asked by Hans Heine of Cologne, Germany)

The term "physiognomy" refers to the features of the face. These features include the shapes and positions of major areas and landmarks of the face such as the forehead, eyebrows, nose, cheeks and mouth. It is a long-standing myth that the human face affects the human personality. The ancient Chinese believed in this myth, as did the ancient Greeks. To the Chinese, the lips were particularly important. For example, in men, the upper lip represented the "feminine side", the lower lip the "masculine side". The upper lip thus revealed how sensual a man was and whether or not he was a good lover. The lower lip indicated how much he needed to be loved. The British notion of keeping "a stiff upper lip" would have been lost on the ancient Chinese!

In the 18th century in Europe, the physiognomy movement, made popular by the Swiss philosopher Lavater, asserted that one's personality and behaviour could be accurately predicted from the face alone. Today we know that the features of the face are genetically determined. The features of the personality are far more variable. In reality, it is impossible to judge someone simply by looking at them — or worse still, looking at a photo. There is no such thing as the so-called stubborn person's chin, hero's cleft, criminal's close-set eyes, jolly person's dimples and so on. However, the face that one shows to the world is reacted to by the world. And the person behind the face reacts back. Therefore, one's face *can* indirectly affect one's personality: for instance, if one has a beautiful face, one will receive positive feedback from other people (who themselves are socially programmed to respond positively to beautiful things as defined by that particular culture). By receiving such positive feedback, one may think more positively about oneself. Hence, one may be more likely to develop a more positive personality generally. The opposite may be true if one has an ugly face. Nevertheless, it is very hard to predict from such data "on the face of it". After all, beauty is in the eye of the beholder.

Although your face doesn't affect your personality, according to Dr Alexander Todorov of the Department of Psychology at Princeton University, "your face speaks volumes about you": "More than just being pretty or handsome, it tells others if you are competent, trustworthy and likeable."[4] In one study, 800 students were shown facial photos of candidates for the US Congress (both winners and losers). It was found that candidates with more mature looks were more frequently picked as "best for the job" than those with a round face, big eyes, small nose, high forehead and small chin (i.e., a baby face). The real surprise was that some of the faces were actually judged to be trustworthy![5]

CAN DOCTORS SWITCH THE FACES OF TWO
PEOPLE LIKE IN THE MOVIE "FACE OFF"?

(Asked by Ben Julian of Marrickville, New South Wales)

The short answer is "No." There are several problems with face transplants, some of which are common to all transplants. First, there is the problem of graft rejection. This is the rejection and destruction of the grafted tissue by the host's immune system. Getting around graft rejection is a hot topic of research, since the only solutions at present involve either finding a donor who is immunologically compatible (which doesn't work in this case, since you're transplanting a specific individual's face), or suppressing the host's immune system sufficiently to prevent the reaction (this being obviously undesirable since it leaves the host open to infection). Other methods such as masking the donor cells or genetically altering the grafted tissue are still being researched, and have not been proven effective in real transplant situations. So let's assume future medical research can find a way around graft rejection. The next problem is nerve connections. There are several nerves that innervate the face, all of which must be correctly connected for the transplant to work. Loss of control or even feeling in the facial muscles causes them to atrophy and droop, altering how the face looks. Nerves are especially problematical, because they cannot be simply sewn back together, no matter how microscopic the surgery is. The problem with nerves is that as soon as they are severed the never fibres (axons) retract away from the cut, and the part not connected to the nerve cell's body (the soma) dies in its sheath. An axon is a long, slender projection of a nerve cell (neuron), which conducts electrical impulses away from the soma. Axons are, in effect, the primary transmission lines of the nervous system.

When surgeons reconnect nerves, they are actually reconnecting the nerve sheaths, so that the nerves can reform their connections without having to make new sheaths (the production of which sometimes prevent the nerves from growing at all). There are so many nerves leading to and from the face that it would take hours of microsurgery to reconnect all of them. And even then the chance of reconnecting them properly would be incredibly slim.

So, if we get the best surgeons to go with our compatible transplant, what else could go wrong? The next major hurdle is with the transplant itself. The shape of the face is not contained in the skin. It is contained in the muscles, bones and cartilage under the skin. So a face transplant would require transplanting the jaw and the entire front half of the skull, being careful not to damage the eyes, olfactory bulbs or frontal lobes of the brain as the bones that support them are removed to their new head. And if this isn't enough, assuming the best and most antiseptic surgical techniques, there is still the problem of attaching the front of one skull onto the back of another, which would assuredly have enough differences in proportion to prevent a seamless attachment. Yet another problem similar to that addressed above is graft versus host disease. This sometimes occurs when bone marrow is transplanted. The donor's immune system attacks the host's body.

However, it is true that by shaving bones and using implants, it is possible to significantly alter a person's physiognomy. Unfortunately, it is almost impossible to duplicate a person's face surgically. After any severe facial trauma, plastic surgeons try to reconstruct the face based on pictures of the person before the injury. Nevertheless, even the best attempts at reconstruction never look exactly right because of small changes in various proportions, shapes and shadows that humans use unconsciously to recognise faces. These probably won't be apparent to someone unfamiliar with the person. But any close friend or relative of the person whose face is being rebuilt from a photograph would be able to distinguish the original from the reconstruction.

DOES THE SHAPE OF MY HEAD DETERMINE HOW I SNORE?

Snoring is at least somewhat related to the shape of your head. If your head is more round-shaped than average, then you are more likely to be a snorer than if your head is longer and more thin-shaped. This is the finding reported by Dr Mark Hans[6] and 5 other researchers in 2001. Their 2-part study yielded the conclusion that head shape can predict snoring.[7] In the first part of the study, the researchers examined the craniofacial characteristics of 60 known snorers and compared their features with 60 individuals with little history of snoring. The researchers examined 25 different parts of the face. They measured the distances from the front teeth to the oesophagus, from the tip of the nose to the rear of the nasal passage, and from the top of the cheekbone to the bottom of the jaw. These measurements formed the basis for a head and face (craniofacial) risk index (CRI). After adding extra factors including age, body mass index and 14 head (cephalometric) measures, the researchers had a useful tool for measuring heads. In the second part of the study, a member of the research team, unaware of the individual's history of snoring, examined the facial features of 19 heavy snorers and 47 light- or non-snorers. Using their new CRI, the researchers tested the hypothesis that head shape could predict snoring behaviour. As it turned out, they were largely right. For at least 75 per cent of the time, it was possible to predict whether the individual was a snorer or not from the shape of their head.[8]

CAN SEX CAUSE A HEADACHE?

It certainly can and it certainly does. Benign coital headache syndrome is also known as orgasmic cephalgia, orgasmic headache or simply sexual headache. Brain specialists have recognised this condition for many years, although they do not know why sex causes headaches in some and not in others. The syndrome affects both men and women. So it's not just "Not now, Harry, I have a headache!"; it could be Harry who has a headache. In fact, British research suggests that men who suffer benign coital headache syndrome outnumber women by a ratio of 3 to 1. The incidence is approximately 1 in every 100 adults. The syndrome seems to occur among people aged 20 to 25 and 35 to 45. It is a mystery why the 26 to 34 age group seems to be less susceptible.

Typically, during sexual activity and just before coitus, the person experiences an intense headache. Headaches from sex have been described in many ways; they have been described as "ice-pick-like" and "explosive", and a Brazilian team of neurologists headed by Dr M. Valenca even described a sex headache as "resembling a thunderclap".[9, 10] However, experts in this area maintain that the headaches at the moment of orgasm differ from both migraines and tension headaches. Nevertheless, a proneness to migraine or tension headaches can predict proneness to benign coital headache syndrome. Dr John Ostergaard of the University Hospital in Aarhus, Denmark, contends that if a patient with migraine or tension headache has 1 episode of benign coital headache, he or she is at great risk of having recurrent attacks.[11, 12] The good news is that benign coital headache syndrome is benign; it is not life-threatening. It is avoidable too — if you don't mind giving up sex.[13, 14]

WHAT IS THE "SOFT SPOT" AT THE TOP OF A BABY'S SKULL AND WHEN DOES IT CLOSE?

(Asked by Scott Harrison of Boston, Massachusetts, USA)

This is one of those OBQs that never goes away. The name for this so-called "soft spot" is the fontanelle. In fact, there are several of them located at the points where the various growing plates of the skull meet.

The anterior fontanelle is located on the top of the head where the coronal suture and the sagittal suture meet. It is the biggest fontanelle at about 2.1 cm in size and often enlarges somewhat in the first few months of life. It also stays open for the longest time, normally closing in the child's first or second year, with 13.8 months being average. It closes by 3 months in about 1 per cent of babies and by 24 months in about 96 per cent of babies. In a tiny number of people it stays open all their lives.

The posterior fontanelle is located at the back of the head where the lambdoidal suture and the sagittal suture meet. It is only 0.5 to 0.7 cm in size and stays open for just 2 months or less. Since it is much smaller and disappears much sooner than the anterior fontanelle, the posterior fontanelle is less well known.

The origin of the word "fontanelle" is interesting too. Fontanelle means "little fountain". It received this name because some say that by looking at the fontanelle, the brain can be seen pulsating, reminiscent of the water rising in a fountain. As I said in *The Odd Body*,[15] you cannot put your thumb through a baby's fontanelle because the baby's skull, although not fully developed, is covered by thick tissue. But don't go trying anything stupid with a baby just to test the point.[16]

WHEN I SHAKE MY HEAD, WHY DOES MY BRAIN MOVE INSIDE MY SKULL?

(Asked by Scott Harrison of Boston, Massachusetts, USA)

Rest assured that even if you shake your head very hard, you will at most move your brain very little. The brain is rather soft and spongy. It is protected within the skull by 3 layers of soft connective tissue called the meninges. When these layers are inflamed due to illness it is called meningitis.

The outer layer is called the dura mater (Latin for "tough mother"). The dura mater is attached to the inner layer of the skull. Just under the dura mater is the second layer, the arachnoid membrane, which looks like a spider's web. There is only the tiniest space between the dura mater and the arachnoid membrane. But if the blood vessels that pass through the dura mater tear (such as in a skull fracture), blood can collect in the tiny space and form what is called a subdural haematoma. The build-up of fluid in this space can cause problems with brain function by squeezing the brain cells. The third layer is the pia mater (Latin for "gentle mother"). This is a thin layer of cells that lines the surface of the brain. Along the pia mater run many blood vessels that eventually enter the brain to supply it with blood. The pia mater is separated from the arachnoid membrane by a small fluid-filled zone called the subarachnoid space. The fluid is called cerebrospinal fluid. This is why it is said that the brain "floats" inside the skull in this very thin zone of cerebrospinal fluid.

So with all of this insulation and protection, it is no wonder that the brain doesn't move very much when you shake your head. Nevertheless, don't experiment to see how far you can go before becoming dizzy, getting a headache or hurting your neck.

WHY DOES NODDING YOUR HEAD MEAN "YES" AND SHAKING YOUR HEAD MEAN "NO"?

A good rule of thumb in science and in life generally is "Never assume". There are parts of the world where the reverse is true — nodding your head means "No" and shaking your head means "Yes". But it seems that the Western custom of nod = "Yes" and shake = "No" is more common. Theory has it that the nodding of the head as "Yes" and the side-to-side shaking of the head as "No" reflect what we did as babies. When a baby wants to cooperate — say, when it wants to eat a certain food that is offered — it will move its head up and down. That movement means "Yes". But when it doesn't want to cooperate — say, when it doesn't want a certain food that is offered — it will turn its head from side to side in an effort to avoid the food. That movement means "No". Throughout our lives we never forget this way we spoke our minds before we could talk.

WHY DO SOME PEOPLE'S HEADS LOOK LIKE THEY'VE BEEN PUSHED OUT OF SHAPE?

(Asked by J. Gardner of New Rochelle, New York, USA)

Obvious causes are injury or a birth defect. One cause of a misshapen head is what's called plagiocephaly, which means "oblique head". Deformational plagiocephaly (aka positional plagiocephaly) refers to a misshapen or asymmetrical head. It's the most common craniofacial problem today. This skull deformity results from repeated external pressure to an infant's skull due to the head's being in one position for extended periods of time. In deformational plagiocephaly, the back part of the head (occiput) is most often flattened. When viewed from above, the head will take on a parallelogram shape and the forehead on the affected side is typically pushed out. The ear on the affected side may be thrust forward compared with the other side. There can be facial asymmetry too, with the affected side having a somewhat fuller cheek. One eye socket, eyebrow or eyelid may not line up very well with the other. According to Dr John Meara,[17] "Infants who spend long periods on their back (e.g., sleeping, in a car seat or pram, or whilst awake) are at increased risk of developing deformational plagiocephaly. Infants should be placed in different positions when awake to avoid constant pressure on one part of the head." A second cause of a misshapen head is craniosynostosis. In this, the infant's skull deformity results from premature fusion of 1 or more of the sutures between the skull bones. Craniosynostosis means "head suture fusion". A third cause of a misshapen head is premature birth. The skull bones of a premature infant are very soft and malleable following birth. They are therefore more susceptible to any external pressures. A fourth but less frequent cause of a misshapen head is a crowded or constrained intrauterine (womb) environment. Put brashly, a baby's noggin can get squished due to a lack of room caused by multiple births, breech position at birth, a small maternal pelvis or a narrow birth canal.[18]

IS IT POSSIBLE TO BE BORN WITH 2 HEADS AND SURVIVE?

It is indeed possible to be born with 2 heads. This extremely rare condition is called craniopagus parasiticus, which literally means "heads fused and living off each other". In craniopagus parasiticus, 2 heads are formed early in embryonic development when the embryo splits into identical twins — but only at the head. As of this date only 6 studies in the medical literature deal with the condition. There have been perhaps 10 cases of the condition ever reported.

The most famous case is the so-called "2-Headed Boy of Bengal", who was born in 1783 and famously described by Dr Everard Home. The boy had an extra head of equal size on top of the first head, but upside down. The lower head had a more or less normal body. According to Home's report, the brains were greatly merged, the skull bones were fused, and the head was covered with black hair. When the boy slept, the eyes of his upper head would open. When he first awoke, all 4 eyes moved in the same direction and then each set of eyes moved independently. Surgical techniques at the time ruled out any attempt at operating. The boy died at age 4 of a cobra bite, unrelated to his condition. His skull is preserved at the Hunterian Museum of the Royal College of Surgeons in London.

A more recent case of craniopagus parasiticus is Rebeca Martinez of the Dominican Republic who, in February 2004, at 7 weeks of age, died after surgery to remove her second head. She had initially survived the surgery, but her brain started haemorrhaging uncontrollably. The most recent case is Manar Maged of Egypt who, in February 2005, at 10 months of age, underwent a successful 13-hour operation to remove her second head. That head included the start of a body (a neck) and was capable of smiling and blinking, but not of independent life. The condition was cruelly satirised through the character of Nurse Gollum in an episode of *South Park*.[19]

WHY DOES MY JAW ALWAYS HURT WHEN I TALK?

(Asked by Cathy Gibson of Ryde, New South Wales)

The "Odd" books never attempt to diagnose illnesses. But if someone is in pain after just talking, it may be a temporomandibular disorder (TMD). These disorders were formerly known as temporomandibular joint syndrome or TMJ — it's good they've shortened the name to TMD because it's a jawbreaker just to say "temporomandibular disorders"! About 7 per cent of Australians suffer from TMD. It is far more common in adults, but children can also suffer from it. For reasons unknown, 90 per cent of TMD suffers are women. And as with adults, girls suffer from TMD far more often than boys.

If you have TMD, the pain can be horrific. You cannot open your mouth as wide as you normally would. When you do, there is a grinding or clicking sound, a feeling that your jaw has come unhinged, and pain ranging from mild to excruciating. Furthermore, the jaw muscles become sore, chewing is difficult, and the pain can spread to the facial and neck muscles. If untreated, the pain can persist around the clock. Headaches, toothaches and earaches may also be symptoms of TMD.

Of course, not all jaw pain is caused by TMD, so other conditions have to be ruled out by a physician. And the causes of TMD have themselves been disputed, so that most medical authorities now believe that there is no single cause. Experts point out that the bones, ligaments and muscles of the jaw form a complicated joint that can be adversely affected by many factors. Here are some suggested causes of TMD:

☞ Bruxism — teeth clenching and grinding — can cause muscle spasms. Spasms can cause more spasms and pain results. Bruxism is often caused by emotional stress.

☞ Malocclusion — teeth that don't fit together properly — can throw the jaw out of alignment and eventually result in jaw pain.

☞ "Internal derangement" of the jaw or other orthopaedic problems of the jaw joint — such as arthritis, bone degeneration, injury or developmental disorders — can cause jaw pain.

☞ Bad posture, particularly thrusting the chin forward, can strain the neck muscles and those of the jaw. Strained neck and shoulder muscles can affect the muscles of the jaw and result in pain. Doctors have warned women for years about the dangers of carrying a heavy shoulder bag for long periods on the same shoulder or gripping a phone between the cheek and the shoulder without using hands during long phone conversations.

☞ A blow to the jaw can be the source of pain.

☞ A whiplash injury from a car accident can affect the

jaw and cause pain. Sometimes this pain can occur a month or more after the accident.

☞ Chewing too many chewy foods or too much chewing gum can overtax the jaw and bring on pain.

Swedish researchers suggest that people who suffer from TMD may have slightly altered brain chemistry. They generally produce lower levels of serotonin. According to Dr M. Ernberg and 4 colleagues, it was also found that TMD sufferers with the lowest serotonin levels experienced the most pain.[20, 21]

The treatment of TMD is varied. Patients who get treatment almost always experience "significant pain relief", according to Dr G.B. Wexler and Dr M.W. McKinney.[22, 23] Should someone experience jaw pain after just talking, they should see their family physician. If it is TMD, the doctor will more than likely say that the first line of treatment is simple self-care. This might consist of going on a soft diet for a few days up to a few weeks and giving up all chewy foods (as well as chewing gum). The doctor may suggest painkillers to ease the pain over this period. Even muscle relaxants may be prescribed if painkillers fail to break the pain–spasm–pain–spasm cycle. Alternate cold and hot compresses applied to the jaw may help too. Doctors often recommend that TMD patients experiment to see which is best for them, and normally recommend resting the jaw as much as possible. Correction of any poor postural or other habit that might be contributing to the problem may be suggested. Finally, a doctor may recommend gentle exercises to relax neck muscles, or may refer the patient to a physiotherapist. It has also been found to be helpful in many TMD cases to squelch any big yawns. You do this by holding the jaw closed with the fingers. Some people, with or without TMD, have been known to dislocate their jaw during a cavernously large yawn.

It's not just a doctor and physiotherapist who'll help a patient overcome TMD. A dentist may be able to correct a problem of poorly aligned teeth that interfere with a correct bite. Having a dentist grind down a few tooth surfaces may be all that's needed. Bite plates or splints fitted over the biting surface of teeth can also help stabilise the bite and eliminate bruxism.

Since stress seems to be related to at least some cases of TMD, pinpointing the source of stress or unhappiness in life may also be important to eliminating TMD. Some kind of psychological counselling may be worthwhile. Talk therapy may help — even if talking hurts.[24, 25]

WHY DOES THE SMILE OF THE "MONA LISA" SEEM TO APPEAR AND THEN DISAPPEAR?

(Asked by Kerry Waterman of Moorebank, New South Wales)

People have been gazing in wonderment at Leonardo da Vinci's *Mona Lisa* for 500 years. First she's smiling, then she's not, then she's smiling again. We read so much into that famous smile and often let our imaginations get the better of us. Philosophers ponder, poets wonder. So much has been written about this awe-inspiring smile.

As for the science, according to Harvard University neuroscientist Dr Margaret Livingstone, the *Mona Lisa*'s smile comes and goes not because the expression is ambiguous, but because of how the human visual system is designed. In staring at the picture, Dr Livingstone claims, there is a kind of "flickering" quality. The flickering is due to the shadows on the *Mona Lisa*'s face as perceived by the eye and occurs as the viewer moves their eyes around the *Mona Lisa*'s face. As Dr Livingstone explains, the smile comes and goes as a result of where the viewer's eyes happen to be. When people look at a face — any face, not only the *Mona Lisa*'s — their eyes spend most of the time focused on the other person's eyes. Thus, when a person's gaze is on *Mona*

Lisa's eyes, their less accurate peripheral vision is on her mouth. Peripheral vision does not pick up detail very well, but it does pick up shadows from *Mona Lisa*'s cheekbones. Dr Livingstone adds that these shadows suggest and enhance the curvature of her smile. But when the viewer's eyes go directly to *Mona Lisa*'s mouth, their central vision does not see the shadows — hence the continuous "flickering" off and on.[26]

A far less complicated view is presented by Dutch researchers from the University of Amsterdam. Dr Nicu Sebe, head of the research team, says the reason why the *Mona Lisa* appears to be mostly smiling is that she's mostly happy.[27–29]

IS IT POSSIBLE TO "READ" A PERSON'S FACE?

More and more experts seem to think you can. The modern science of reading facial expressions owes a great deal to Dr Paul Ekman, a psychologist at the University of California Medical Center in San Francisco. Ekman studied the human face — its features and expressions — and its relationship to human behaviour. His goal was to be able to predict human behaviour via clues indicated by the face. His research led to a detailed set of rules for reading and interpreting faces and resulted in the formulation of the Facial Action Coding System (FACS). In FACS, Ekman classified 3000 facial expressions and meticulously studied when each is used in human interaction. Associations between facial expressions and subsequent behaviours were revealed.[30]

FACS has since been used by a variety of researchers and by many others: legal researchers have used it to study jury psychology; animators have used it for movies such as *Toy Story* and *Shrek*; and the CIA and the FBI have used it to assist in counter-terrorism training. The theory behind the latter is that face-reading skills can help interrogators better determine whether or not there are any discrepancies between what a suspect says and what is signalled by their face. Ekman points to several ways of establishing the truth through observing facial expression. For example, a fake expression indicating that a person is lying is generally held longer than a real expression. In lying, there is usually more blinking, more looking away, and more smiling with only the mouth. In a true smile, the mouth is more consistent and coordinated with the eyes, in what most of us would recognise as an "open" smile.

Although it seems to be possible to "read" a person's face, it is an inexact science. A skilled and practised liar can often still get away with it, as can a true psychopath.[31]

WHY DO THEY CALL PLASTIC SURGERY "PLASTIC"?

(Asked by Mark Lewis of Storrs, Connecticut, USA)

It is not widely known, but the surgical speciality of plastic surgery seldom if ever uses any plastic. In fact, the name actually has no connection at all with synthetic polymers (i.e., plastics). In this sense, plastic derives from the Greek word *plastikos* meaning "able to be moulded". Plastic surgery deals with the appearance, form and surgical reconstruction of body tissues — not the use of plastic substances to do so. Plastic surgery is also much older than most people think. The ancient Egyptians, the early Greeks, the Hindus of India and physicians over the past few centuries have attempted to change facial characteristics through surgery of various sorts. However, the American Society of Plastic Surgeons was only formed in 1931, and modern techniques did not develop until after World War I, when surgeons attempted to repair the disfigurement of combat. This marks the beginning of the true profession of plastic surgery. Breakthroughs in plastic surgery have occurred particularly in the last 20 years.

It is popularly believed any surgeon can become a plastic surgeon. This is not true: plastic surgeons require as much or more postgraduate training than any other surgical specialist. It is also believed that plastic surgery is a frivolous speciality concerned only with cosmetic facelifts, breast enlargements, tummy tucks and such. This is also not true: more than 60 per cent of all plastic surgery deals with reconstruction to repair serious damage from burns, injuries or congenital abnormalities. Less than 40 per cent involves cosmetic surgery for reasons of vanity only. Of this 40 per cent, in 2002 in just the US, 6.6 million people had cosmetic plastic surgery of one kind or another. Eighty-five per cent of these were women, 15 per cent were men. The top 5 'cosmetic surgical procedures' for women were: (1) breast augmentation, (2) liposuction, (3) nose reshaping, (4) eyelid surgery and (5) facelift. The top 5 "cosmetic surgical procedures" for men were: (1) nose reshaping, (2) liposuction, (3) eyelid surgery, (4) hair transplantation and (5) ear surgery. Of those having cosmetic plastic surgery 45 per cent were aged 35 to 50, 24 per cent were aged 19 to 34, 22 per cent were aged 51 to 64, 6 per cent were aged 65 and over, while 3 per cent were aged 18 and younger.[32]

WILL WE EVER DEVELOP AN ARTIFICIAL FACE?

(Asked by Ronnie Gilbert of Great Neck, New York, USA)

Most experts agree that a complete artificial face is still some years away. However, today the chin, jaw, lips and other parts of the face can be largely rebuilt through silicone implants. Cranial–facial surgery has made great strides in recent years. This form of surgery is performed to correct a misshapen jaw or facial asymmetry due to a congenital abnormality or a serious illness or injury. Malarplasty is the medical term to describe the augmentation or general reshaping of the cheeks and jaw. In malarplasty, implants are used to change the underlying cheek and jaw structure. This affects the overall balance of the face. The implants come in a variety of shapes and sizes. They are made of both solid and semi-solid materials. Cheek augmentation involves placing the implant over the cheekbone through an incision made inside the mouth or through an opening just beneath the lower eyelash. The surgeon creates a pocket in the tissue and then inserts the implant. The incision blends with the lash line and is nearly unnoticeable. It sounds awful and painful, but the surgery takes only between 30 and 60 minutes and in many circumstances may be done under just a local anaesthetic.[33]

Polygraph lie detectors are notoriously unreliable. Estimates are that 20 per cent of those telling the truth fail the test, while 10 per cent of liars pass the test. Computer programs that measure the tiniest changes in facial expressions when people are testifying have been developed. One is from the Institute for Neural Computation in San Diego. Inventors of the program claim that if lying involves subtle changes in facial expressions, they are able to identify liars using their program far better than using a polygraph.[34, 35]

It has been determined that one brow wrinkle is the result of 200,000 frowns.

According to the American Society of Plastic Surgeons, migraine headache sufferers can get significant relief or elimination of symptoms through modifications of plastic surgery procedures traditionally used to minimise facial wrinkles.[32]

CHAPTER 3

THE EYES

The eyes figure in the proverbs and wise sayings of many peoples throughout the world. A Jewish proverb says, "What you don't see with your eyes, don't invent with your tongue." A German one says, "The eyes believe themselves; the ears believe other people." And a French one goes "Love is blind; friendship closes its eyes." An ancient Greek proverb, often attributed to Plato, observes, "The spiritual eyesight improves as the physical eyesight declines." And there's so much more about eyes — as you can see!

DO MY EYEBALLS REMAIN THE SAME SIZE FROM BIRTH?

(Asked by Jason Janus of Bellevue Hill, New South Wales)

Not quite. The diameter of a newborn baby's eyeball is about 18 millimetres. It increases rapidly in size during the first year of life. But growth is very slow after that. According to the classic medical text on eyes, *Ocular Pathology* by Dr David J. Apple and the late Dr Maurice F. Rabb,[1,2] the reasons behind eyeball growth are not precisely known. In animal experiments in the 1950s, it was found that eye growth in a developing chick can be diminished by introducing a small tube into the eye to drain off intraocular fluid and reduce intraocular pressure. When this is done, the eyeball fails to enlarge normally. But strangely the retina keeps on growing as if nothing happened. By the time the chick is ready to hatch from the egg, the eyeball is smaller than normal, but the retina has enlarged to form a folded sheet of neural tissue crowded within the eyeball. Poor little chick! Animals such as fish are apparently able to enlarge their eyeballs and retinas steadily over the course of their entire lives. However, this ability is lost in birds and mammals — including humans.[3–5]

WHAT AM I LOOKING AT WHEN I SHUT MY EYES?

(Asked by Jason Janus of Bellevue Hill, New South Wales)

When your eyes are shut, you are looking at the insides of your eyelids. However, you do not see anything in much detail. This is because there just isn't enough light to illuminate anything, unless the place you are in is very bright. Nevertheless, you may notice a reddish glow to any light that gets through to your retina, since it has to pass through the eyelid. According to *Adler's Physiology of the Eye*,[6] the reddish glow is due to the fact that your eyelid is filled with blood vessels. The retina does not automatically "turn off" even though your eyelids are shut. Any object you could see would be out of focus in any case, since the eyelid, located just next to the surface of the eye, is beyond the eye's range of focus.[7]

CAN MY EYES CHANGE COLOUR?

(Asked by Paul Casey of Leeds, UK)

It surprises most people to learn that eye colour *can* change. But it rarely happens. There are some conditions and diseases of the eyes that can cause the coloured part of the eye (the iris) to change colour. This usually affects only one eye. But if both eyes are affected, usually one eye is more affected than the other. Some infants who are born with very light-coloured eyes will have a change in eye colour to green or even brown as their eyes mature normally. It's also possible for a person to be born with a mole on their iris. The mole, being brown in colour, when covering an iris that is blue or grey, makes the eye look brown or even black. Eye colour is governed by genetics. It's fairly common for someone to be born with one eye in one colour and the other eye in another colour. But whatever that colour, chances are it will stay that way.[7]

ARE LIGHT-COLOURED EYES MORE SENSITIVE TO LIGHT THAN DARK-COLOURED EYES?

(Asked by Paul Casey of Leeds, UK)

Humans have different eye colours depending upon the amount of pigment found in the iris of each eye. Brown eyes have more pigment than green. Green eyes have more than blue. Some people are born with little or no pigment. The pigment acts as a barrier or protection for the structures behind it. The more pigment, the better the protection from harmful UV light that can damage the lens and the retina. Cataracts and macular degeneration are some of the more common complications of such damage. It is not clear whether light-coloured eyes are necessarily more "sensitive" to light than darker eyes. This is because "sensitive" to one person is not "sensitive" to another.

While someone with very dark brown eyes has more pigment in their iris tissues, which will be a more effective shade as their pupils get smaller, we can adjust our light sensitivity by means of our retinas as well. According to Dr Ken Mitton of the Department of Ophthalmology at the University of Michigan in Ann Arbor, the photoreceptor cells in our retinas allow us to see by converting photons of light into nerve impulses that will travel along the optic nerves to our brains. The levels of the chemical reactions in the photoreceptor cells and the numerous interacting proteins and chemicals involved are adjusted in different lighting conditions. In all humans, rod cells provide black and white (grey-scale) information, while red, green and blue cone cells detect the red, green and blue light ranges. All of these separate detectors give us the final colour image that we see in our "mind's eye". In bright sunlight, only our cone cells (colour) are working — they work well in bright light — while our rod cells are "bleached" and are not providing any information at all. In the dark, our very sensitive rod cells (black and white) give us the picture, while our cone cells do not get enough light to detect anything at all.

WHY CAN'T HUMANS SEE INFRARED LIGHT?

(Asked by Martin Langford of Peakhurst, New South Wales)

This is an intriguing OBQ. According to Dr Tom Stickel,[8] the human eye is capable of detecting electromagnetic radiation in a range from about 380 nanometres (nm) of wavelength to 700 nm. The radiation between 380 and 700 nm is thus called visible light. Wavelengths at around 400 nm appear violet. Therefore, wavelengths much shorter than 400 nm are in the "ultra" violet (UV) range, and are out of the range of detection by the human eye. Wavelengths around 700 nm appear red. Beyond 700 nm, the radiation is referred to as "infra" red, or IR. Anything that gives off heat is giving off some form of infrared radiation.

The reason why we can see colours at all is because we have 3 separate proteins in the back of the eye, each specialising in a specific wavelength. These proteins reside in cells called cones. Each cone has only 1 type of colour protein in it. At the 380 nm wavelength the S cone ("S" for short wavelength) is the only cone detecting the light. This is because 400 nm is out of the range of the other 2 cones. Similarly, at 700 nm, only the L ("L" for long wavelength) cone is operating. This is because the other 2 cones don't detect light of that wavelength. Natural variation being what it is, some people will probably have slight variations in their L cones that give them better colour vision than others in the range of red colours. Thus, it seems likely that some people will be able to see longer wavelengths than others. Similarly, some people can see shorter wavelengths than others. Either way you look at it (no pun intended), as long as it can be seen, it's visible light and therefore not IR — or UV on the short wavelength side.

Physically, there is very little difference between 700 nm light and 725 nm light. Nevertheless, humans can detect 1 and not the other. There are arbitrary cutoffs for where visible reds end and near-IR begins. As stated above, the range of visible light is between about 380 and 700 nm. It is "about" because there is variation between people at the ends of the ranges. Again, as long as it can be seen, it's not technically IR, it's still red. In any case, due to natural variation, some people can probably see a tiny way into the near IR range as given by some arbitrary definition. But that doesn't give them any special powers or capabilities. It just lets them see reds that are very slightly dimmer than what others can see. It is unknown what the record is for how far into the near-IR range anyone has detected red light, but it's most probably only a few nm further than the population average.[9]

WHAT MAKES ME SEE "SNOW" WHEN I SEE A BRIGHT LIGHT?

This phenomenon may occur when one gazes out the window of an aeroplane on a particularly bright day. Some people see "snow". Others describe it as "sparkles", still others as "stars". In any case, what you are seeing is your own blood cells flowing through translucent blood vessels in front of the eye's retina. Normally we do not notice the pattern of these blood vessels because the brain knows that they are there and compensates for the effect by screening it out of our normal perception. But when exposed to the combination of a very bright light and the continuously changing position of the blood cells circulating in the blood vessels, they become noticeable. In fact, the snow, sparkles or stars we see will appear to move in time with our own pulse. It seems that a bright blue light works best in producing this effect. The phenomenon was once the subject of an exhibit at the New York Hall of Science. Visitors could gaze into a box containing a bright blue light and experience the effect for themselves. Furthermore, it has been suggested that the phenomenon could be developed into a diagnostic tool in which patients with certain blood diseases could estimate their own blood-cell counts. But this might be a bit far-fetched today. Perhaps a "vision" of the future?

WHAT CAUSES A BLACK EYE?

(Asked by Joe Cooper of Ventura, California, USA)

Getting punched! Any boxer and most schoolboys will tell you that. The medical term for a black eye is ecchymosis. What a great term! Sounds like some sort of fatal skin disease. Ecchymosis is caused by bleeding (haemorrhaging) around the eye. A blow causes damage to blood vessels. The colour of a black eye changes because of the changes in the haemoglobin molecule in the blood. Haemoglobin is red in its normal state of carrying oxygen. In fact, haemoglobin is perhaps best described as a large, complex pigment molecule. When trauma occurs to the skin around the eye, the liquid portion of blood (plasma) is quickly reabsorbed into the bloodstream and surrounding tissues. However, the rest of the blood cells are torn apart and their contents are released into those tissues. This blood cell debris consists almost entirely of haemoglobin molecules. As the black eye heals you see the haemoglobin change under the skin, first appearing as black, then purple, then blue, then yellow, and finally red again. A hyphema is when bleeding occurs within the eye. A black eye can hide a skull fracture and should be reported to a doctor. Watch out for swinging doors!

WHY DO MY EYES GET BLOODSHOT?

(Asked by Richard Jenkins of Bendigo, Victoria)

Bloodshot eyes are caused by stress to the surface of the eyeball. Something is irritating them. Many different things can irritate your eyes. You can get a foreign substance in them, stare at a computer screen for several hours, let your eyes dry out and then rub them, or get punched in one. All of these things will make your eyes bloodshot. Your eyes are full of little blood vessels. Look at the whites of your eyes really closely in the mirror and you will see blood vessels over the entire surface. What happens when your eyes get irritated is that the blood vessels swell up and start to look big and red. Redness, heat, swelling and pain are part of the inflammation process, which is a way of repairing the body. You get a very similar response any time something irritates any part of your body. It's just that you can't always see it.

HOW LONG DOES IT TAKE TO DAMAGE YOUR EYES WHEN YOU STARE AT THE SUN?

(Asked by Tommy Wallace of Pennant Hills, New South Wales)

The simple answer to this is to never stare at the sun — not for a second! It's too dangerous! Don't even think about testing this out. Never staring at the sun is the only way to be absolutely sure your eyesight won't be in danger. This is because looking at the sun can instantly damage your eyes. Moreover, the eye's retina has no pain receptors. So, it's very easy to stare at the sun too long and not realise that eye damage has occurred. Apart from this, the actual time that it takes to damage the eyes by looking at the sun ranges from instantly to a very few seconds to more than a very few seconds. The time depends upon several factors, such as the brightness of the sun on a particular day (influenced by clouds and the height of the sun in the sky), whether or not the sun is stared at with the naked eye or by using a telescope or binoculars, the health of the eye, and so on.

The sun emits both infrared and ultraviolet radiation. When you stare at the sun, both forms of radiation are focused and concentrated onto your retina. In bright sunlight, the cells at that point on your retina could be damaged beyond repair in a few seconds. In addition, your retina contains a region of very sensitive cells called the yellow spot or fovea. This is responsible for the ability of the eye to detect fine details in only dim light. The destruction of these cells causes substantial visual impairment. The sun can destroy these cells faster than in the time it takes to read this sentence. As your mother said, "You only have one set of eyes." And as she also said, "It's not polite to stare." But if you must, stare at other things — never at the sun.[10]

WHY DO I SOMETIMES VISUALISE A GLOWING GREEN OR YELLOW "EYE" AFTER RUBBING MY EYES?

(Asked by Anna Jeffes of Walnut Creek, California, USA)

According to neurosurgeon Dr John Morenski,[11] when you apply pressure to an organ you can actually transfer energy to that organ. This is what happens in an emergency room of a hospital when a "precordial thump" is administered. A medical staff member pounds the chest of a patient who is in ventricular. This is a malignant heart rhythm that can rapidly lead to death. The energy transferred by the "thump" can in some cases clear the heart of its malignant rhythm. As for rubbing the eye, the pressure of a rub is sufficient to trigger your photoreceptors. These receptors and the brain itself have no reason to interpret the signal as anything other than a real object in the visual field. Just how the eye lets you see, and exactly how the images make it to your brain, as well as how the brain interprets the message, prove to be a very complicated process.[12]

WHY DO I GET EYESTRAIN?

(Asked by Tommy Wallace of Pennant Hills, New South Wales)

Maybe Grandma *wasn't* right. Experts now contend that reading in the dark probably has no long-term damaging effects on the eyes. But eyes become more tired in poor light, since they have to work harder to discriminate between letters, colours and objects. So it's a good idea to work and play in plenty of light, especially as one gets older. It's a sad fact with eyes that as one gets older, pupils get smaller. When you're older and wish to read with ease, it may take a little more light to illuminate the retina to the same level as it took to read with ease as a youth. According to Chicago optometrist Dr Kenton L. McWilliams, expert opinion is divided as to whether or not regular eye exercises can strengthen the muscles in the eyes that are responsible for focusing. Eye exercises, also known as vision therapy, should only be undertaken on the advice and monitoring of an ophthalmologist or optometrist. Such vision therapy is usually intended for younger patients who have undeveloped or underdeveloped visual problems that may or may not involve specific eye muscles.

Some problems are perceptual in nature. The problem in middle age, when focusing on close objects becomes a strain, affects everybody at different times, depending on a host of factors. Some major factors include: refractive status of eyes, e.g. nearsightedness (myopia) and other conditions; daily visual demands; general health status; medications taken; and so on. The muscles responsible for focusing do not really change. It is the transparent lens that actually gets less malleable with age. The chemical properties of the lens of the eye change and harden and the lens is prohibited from bending and changing thickness as easily. Human evolution is probably to blame for this, as our prehistoric ancestors probably did not depend on eyes for close work nearly as much as we do.

DO BLIND PEOPLE REALLY HEAR BETTER?

(Asked by Candace Bollinger of Glebe, New South Wales)

This is another classic OBQ. Blind people *do* hear better than those with normal hearing. Specifically, blind people are better than sighted people at discerning the direction from which sounds come and also changes in the pitch of sounds. This is especially true for those who become blind early in life, or "the early blind". According to Dr N. Lessard and colleagues from the Sacred Heart Hospital in Montreal, a famous example of someone who adapted their senses is Helen Keller (1880–1968). Blind and deaf from an early age, Helen Keller developed her sense of smell so finely that she could identify friends by their personal odours as soon as they entered the room she was in. The superior hearing ability of the early blind probably has to do with the fact that after it loses its original function, the part of the brain most responsible for vision (the occipital lobe) starts helping out in other functions such as hearing.[13, 14] Experiments also reveal that blind people are much better than sighted people at orienting themselves by sound. A sighted person who is stuck in traffic and hears a fire engine will look around to see where the sound is coming from. But a blind person is better able to pinpoint the sound. This theory has been given support recently by Dr Pascal Belin and colleagues from the Department of Psychology at the University of Montreal.[15] This is yet another example of recent findings in brain research showing the brain to be more able to change in response to injury or illness than researchers thought even a decade ago. And it helps explain why Stevie Wonder (blind since birth) the late Ray Charles (blind since age 7), Andrea Bocelli (blind since birth) and many others display remarkable musical gifts.

DOES PHYSICAL ATTRACTIVENESS AFFECT A JURY?

The verdict is in. Physical attractiveness *does* affect a jury. What the jury sees makes all the difference in the world. Research shows that physically attractive people enjoy many advantages over unattractive people in courts of law. It is called the attractiveness-leniency effect. For example, there is a greater likelihood that a physically attractive defendant will be acquitted of a crime, or, if convicted, will receive a lighter sentence.

In 1 study, it was found that characteristics of attractiveness include "a friendly, self-assured expression, stylish hair, and a well-proportioned body". Those considered to have characteristics of unattractiveness had "an unrefined appearance, dress unfashionably or informally, round face, and stout body". In the study, the attractiveness-leniency effect was shown to be "a significant extralegal factor" in decisions of juries in Canada, Taiwan and the US.[16] The authors of the study, headed by Dr Karl Wuensch of East Carolina University, point out that the notion of "what is beautiful is good" in jury thinking has been known among specialists in jury psychology since 1972. Studies since then have shown that juries are swayed one way or the other by the physical attractiveness of defendants, plaintiffs and witnesses. This is often completely independent of the actual facts of the case. The same research team has also found that physical attractiveness is a factor in sexual harassment lawsuits as well.[17]

Gender and race may be factors too. In 1 experiment, Dr D. Abwender and Dr K. Hough[18] write that "After reading a vehicular-homicide vignette in which the defendant's attractiveness and race varied, the participants [in the experiment] rated guilt and recommended sentences. The women treated the unattractive female defendant more harshly than they treated the attractive female defendant; the men showed an opposite tendency [they treated the attractive female defendant more harshly, the unattractive female defendant less harshly]. The Black participants showed greater leniency when the defendant was described as Black rather than White. The Hispanic participants showed an opposite trend [showing less leniency when the defendant was described as Hispanic rather than White], and the White participants showed no race-based leniency."[19]

The average human eyeball weighs about 28 grams (1 ounce).[12]

About 94 per cent of women and 50 per cent of men break down in tears once a month.

A healthy newborn baby has 20/50 vision. This means that what the average person can see at 50 metres, the newborn can only see at 20 metres.

Humans blink every 2 to 10 seconds.[12]

The pupil of the eye expands as much as 45 per cent when a person looks at something pleasing.[12]

Most of the human behaviour of flirting is done with the eyes. A psychologist reports in a study of flirtation that women start the flirtation 2 out of 3 times. Mostly this is done with non-verbal cues in the categories of the 4 "Gs": gazing (using the eyes), glancing (using the eyes), giggling and gesturing. There are 52 flirtation cues in all.[12]

CAN YOUR EYES BE USED TO IDENTIFY YOU?

They certainly can. As early as 1935, researchers put forward the view that the eye could be used for identification. Our eye, like our fingerprint, is unique to each and every one of us. There are 2 ways of identifying you through your eyes: iris identification and retinal identification.

Iris identification

No 2 irises are alike, even in identical twins. The iris is a consistent means of identification, since it remains unchanged throughout life and is not subject to wear or injury. In the iris there are 400 distinguishing characteristics that can be counted and used to identify someone. These are called "degrees of freedom". The iris has 6 times more distinct identifiable features than a fingerprint. The chances of finding any 2 that are the same are 10 to the 78th power. The entire population of the Earth is about 10 to the 10th power. There are 2 methods used in iris identification systems: active and passive. The active iris system requires the subject to move back and forth so that a camera can adjust itself and focus on the subject's iris. This requires the subject to be between 15 and 35 centimetres away from the camera. The passive iris system uses a series of cameras that locate and focus on the iris. The subject may be as far away as 1 metre from the cameras. With either system, it takes about 2 seconds to make a positive identification.

Retinal identification

The pattern of blood vessels on the back of the retina is also unique to each of us. Retinal scans involve a low-intensity infrared light that is projected through to the back of the eye and onto the retina. Infrared light is used because it is absorbed faster by the blood vessels on the retina than by the surrounding eye tissue. The infrared light with the retinal pattern is reflected back to a video camera that captures the retinal pattern and makes the identification. Retinal identification has several disadvantages, however: the retina is susceptible to conditions such as cataracts; the technology is somewhat intrusive, as the subject must place their eye near the camera; and a trained expert must supervise the procedure. Yet when the system is applied correctly there is nearly a zero error rate.[20, 21]

HOW DOES PERIPHERAL VISION WORK?

Our peripheral vision works in essentially the same way as our central vision. The major difference is the quality (or acuity) of what we see. According to Dr Sarah McKay, a neurologist from the Department of Physiology at Oxford University, in our retina there are 2 classes of light receptor cells that detect the light we see. Both cell types can be found in the centre of our retina (the fovea) and the periphery of the retina, but there are differences in the distribution of these. Clustered around the fovea are cones, the cell types that pick up different wavelengths of the light spectrum. They are very densely packed here and thus able to detect very fine details. They are much less densely packed in the periphery. The other cell type is known as rods. These are more densely packed in the periphery than in the fovea. On a normal sunny day these don't do much work, as they are used for detecting very dim light and are the ones that kick in when night falls. One fun way of using your rods is when you are looking up at stars in the sky. You might have noticed that you can see very faint stars much better if you don't look directly at them. In this case you are letting the dim starlight fall on the periphery of your retina rather than the fovea. Thus the rods are being stimulated instead of the cones (which aren't working too well, since it is night-time), so you see the star!

Signals from your retina are passed onto the rest of the pathways in your visual system, but since the most important part of what you see is what you are focusing on, the majority of the processing is taken up by what falls onto your fovea and thus the cones. So larger parts of your brain are taken up by processing what is in the centre of your vision than what is in the periphery. It might not seem like this, but try focusing on a point in front of you. For example, try this page. Focus on the word "page" you have just read. You won't actually be able to see what the words say even 3 or 4 lines above or below that (i.e., in your periphery).

So this gives you an idea of how important the brain thinks our peripheral vision is. But there's no need to worry. That's why we can move our eyes and our heads![22]

HOW CAN AN EYE LENS BE CLEAR YET CONTAIN BLOOD?

(Asked by Jackie Lance of Lismore, New South Wales)

It is inconsistent to imagine a lens that is both colourless and has red blood flowing through it. So it must be that one of these assumptions is wrong: either the lens is not clear or it doesn't have blood flowing through it. In fact, it is the latter assumption that is wrong. According to Dr Tom Wilson,[23] the lens of the eye does not have blood flowing through it. But how can this be, you ask? The lens is alive, isn't it? And living things need blood to bring them oxygen, don't they? If the brain can't live without blood, how can the lens?

In fact, the lens is living tissue and there are living cells within it. But these are very special cells with special properties. Most importantly for us, they are adapted to survive and grow with very little oxygen. This is in large part because they grow very slowly over the lifetime of the person. Still, they do need *some* oxygen, and they do need to get rid of other waste products. The lens cells must rely entirely on diffusion (i.e., passive exchange) of these materials into and out of the lens. To facilitate this, the lens is bathed in a flowing bath of fluid called the aqueous humour, which is located just in front of the lens and behind the cornea, which covers the surface of the eye. This fluid contains no blood cells, so it is clear. But it can slowly transport gases and other materials into and out of the lens. The fluid is constantly recycled so that these gases and materials don't build up.

Note that this process of diffusional exchange is fundamentally the same as what any cell does in an organ with blood flow. In other words, all cells rely on diffusion of these gases and waste products, it's just that the lens doesn't get the added benefit of nearby, fast-flowing red blood cells to facilitate the diffusion. This means that the lens cells grow and metabolise at a much reduced rate. It also means that the size of the lens is limited. If it got too big, the gases and other materials could never diffuse to the middle of the lens.

ARE HUMANS THE ONLY ANIMALS THAT HAVE EMOTIONS?

Attributing human qualities to animals is called anthropomorphism. It is impossible to know what animals are really thinking and feeling, but there are indications that they experience something very close to what we would recognise as emotions. Some animals are monogamous and mate for life. Is this the emotion of love? Animals emit sounds when they are in distress. Baby animals will vocalise mournful sounds when separated from their mother. For example, the sounds made by bear cubs have been described as remarkably similar to the cry of a human infant. Animals can appear to grieve. An elk will guard its dead calf for days against predators. Dolphins may carry a dead calf for days. The famous Greyfriars Bobby, a folk hero of Scotland and the subject of a Disney movie, was a small dog whose owner passed away, and who then kept constant watch over his owner's grave until his own death 14 years later. Is this a sign of grief? Love? You decide. Or perhaps ask your pet?[24-26]

WHY DO WE SEE OPTICAL ILLUSIONS?

Optical illusions occur because our brain is too good at interpreting what it sees. We live in a 3-dimensional world (up–down, right–left, backwards–forwards). We are accustomed to viewing the depth cues of this 3-dimensional world. When presented with a 2-dimensional drawing, the brain tries to reinterpret the flat image as if it were 3-dimensional. "The brain is not built to look at pictures. That's the basis for many illusions," claims Dr Rudier von der Heydt, a physiologist at Johns Hopkins University in Baltimore. Studies show that you can't refuse to experience an optical illusion. The ability to interpret depth cues or complete broken contours is simply too important to perception to be turned on and off at will. Dr von der Heydt has found evidence that the action of seeking out contours is hard-wired into our brain cells. Other studies show that even infants have some sense of depth and edges and of when something is partially hidden. As Dr von der Heydt adds, "The brain is always striving for a three-dimensional interpretation."[27]

In the Ponzo illusion, several lines run across a page but the top horizontal line appears longer than the bottom line, even though they're really the same length. This is so because the brain is accustomed to observing in real life that the parallel sides of a road appear to converge as the road disappears into the distance. Because the brain is used to reading perspective cues this way, it concludes that the horizontal line in the top part of the Ponzo illusion — where the vertical lines almost converge — is more distant than the line in the bottom part of the picture. But since the "more distant" line doesn't appear smaller than the "nearer" line, the brain mistakenly concludes that it must have been larger in the first place.

In the vase–face illusion there is a figure that could look like either a vase or two faces in profile. Again the brain thinks in 3-D. In the everyday world, our visual surroundings consist of objects set in front of other objects or backgrounds. The brain gets used to this and looks for an object in the foreground and an object in the background. In the vase–face illusion, it sees a vase if it chooses white for the foreground and black for the background. It sees two faces if it chooses black for the foreground and white for the background. We can control the process by a mental "flip-flop" back and forth between the 2 images as we wish.

The Kanizsa triangle illusion, where a series of figures must be interpreted at the same time, illustrates how even looking for the outlines of objects can fool the brain. In trying to distinguish an object from its surroundings, the brain always looks for the object's contours. This is so even when the object is the same colour as its background or when its edges are partially hidden by something in front of it. This ability to infer objects and contours when they don't exist deludes you into seeing 2 triangles and 3 circles.

DO YOU BELIEVE YOU CAN SENSE WHEN SOMEONE IS LOOKING AT YOU EVEN WHEN YOU CAN'T SEE THEM?

If you do, you're not alone. A surprisingly large 87 per cent of us think we know when we're being watched by a hidden observer. But do we really have this ability? Rupert Sheldrake, a London biologist and author, thinks that we do — under the right conditions.[28] But the evidence for this is mixed. Dr Gary Rosenthal and 3 colleagues from Nicholls State University in Louisiana decided to find out for sure. They conducted an experiment to test our ability to detect a hidden observer.[29] In their experiment, 140 subjects individually sat in a room with a 2-way mirror and a video-monitor camera. The subjects were told that they might be observed for any or all of the next 5 minutes. Subjects noted whether or not they felt a hidden observer watching them during each minute of the 5-minute session. After the session, subjects reported on whether they felt a hidden observer had been watching them via either the 2-way mirror or the video, or felt they hadn't been watched at all. The researchers found that subjects were not able to detect when they were being observed beyond what would be expected by chance.

So it seems that we don't have the ability to detect when a hidden observer is looking at us after all. But why do so many of us believe we do? There are 2 theories that could account for this phenomenon. First, a confirmation bias may be at work. A confirmation bias occurs when a person who believes in something seeks out information that supports their belief, but ignores information that refutes it. Second, differential remembering may be at work. Differential remembering occurs when a person more readily recalls information supporting their belief than information not supporting it. In both instances, people don't know that they're thinking this way and end up fooling themselves. The Rosenthal team points out that 87 per cent of their subjects believed they could detect a hidden observer before participating in the experiment. Yet even after participating in the experiment and confronted with proof to the contrary, more than ¾ of subjects claimed they could detect a hidden observer.

Both confirmation bias and differential remembering probably explain many things about human behaviour. They can explain why people are quick to attribute a cause-and-effect relationship to things that are entirely unrelated. And they also explain, to some extent at least, why prejudices persist. For example, one day a person sees a man wearing a turban. They then see the man in the turban steal a piece of fruit from a store and draw the erroneous conclusion that all men who wear turbans are thieves. Even worse, Islamic men are more likely to wear a turban, therefore Islamic men are more likely to be thieves. Such illogic is differential remembering at work. Or perhaps after seeing the man wearing a turban steal a piece of fruit, the following conversation occurs in a person's mind: "I think all men with turbans are thieves. Here is a man with a turban stealing something. Therefore, I was right. All men wearing turbans are thieves. Islamic men wear turbans. Therefore, Islamic men are thieves." This is confirmation bias at work.

WHY DO I SOMETIMES SENSE THINGS THAT AREN'T THERE?

Many of us believe we see and feel things that aren't there. Throughout the centuries, individuals have claimed they have seen gods, angels, spirits, ghosts, demons and extraterrestrial beings of various sorts. Two Canadian researchers, Dr C.M. Cook and Dr M.A. Persinger,[30] decided to test their theory as to why some people believe they have what the 2 scientists call "sensed presence".[31] In a simple experiment, 15 subjects were asked to sit in a room by themselves and press a button whenever they felt a mystical presence was with them. Unbeknownst to the subjects, they were occasionally exposed to a weak magnetic field that was turned on and off at random by the experimenters. The researchers were surprised to find that the feeling of a "mystical presence" indicated by a press of the button coincided with the application of the magnetic field more often than mere chance would predict.

Cook and Persinger were left with 2 possible conclusions. Either subjects were far more sensitive to changes in the magnetic field around them than anyone had previously thought, or subjects did indeed possess "sense presence". Cook and Persinger theorise that any "sensed presence", when it is experienced, is a short-circuit illusion somehow created by the miscommunication between the 2 halves of our brain. The left half of the brain holds information about what truly exists and our "sense of self" (the conscious, reflective personality of the individual). This is absent in the right half of the brain. The illusion is really only a fleeting right-brain reflection of the left-brain's "sense of self". We *think* we see and feel something, but it isn't really there. Moreover, Cook and Persinger claim that this short-circuit illusion probably occurs "along that interconnected conduit [of the brain] called the corpus callosum". The corpus callosum is an information pathway. In "sensed presence", information takes a detour. Such a theory is consistent with the ideas of many brain experts, who explain the mind as merely a creation of the brain.[31]

DO SUPERMARKETS HYPNOTISE YOU TO SHOP?

(Asked by Brad Townsend of New York City, New York, USA)

Supermarkets don't hypnotise you as such, but they do apply some principles of consumer psychology to get you to spend more. When most consumers enter a supermarket, they carry on walking for at least a few seconds before adjusting to their surroundings. So supermarkets usually leave the entrance area clear of stock to help consumers feel comfortable. They will also make sure this part of the supermarket is brightly lit. In cold weather, customers walking through the door are often greeted with a blast of warm air. In hot weather, it's cool air. This makes consumers feel as though they've entered a welcoming environment and quickly puts them in the mood to browse and spend. Fresh fruit and vegetables are placed at the front of the store, even though they might be squashed at the bottom of the trolley since they're bought first, but the psychological association with freshness and quality is so powerful that it has a positive effect on sales. About 75 per cent of customers look to the right when entering a supermarket. Knowing this, supermarkets usually put the best offers on the right. Also on the right, by the front door, is often an area for impulse buys such as magazines, soft drinks, and bakery goods. Sweets and snacks are almost always in the last aisle before the checkout. Consumer psychology reveals that after consumers buy the healthy and important items, they are more likely to buy something unhealthy and unimportant — rather than the other way around.

Consumers "read" shelves like a book. So supermarkets normally put the most expensive items on the right, where a shopper's eyes are likely to linger longer. Also, a shopper's eyes normally look straight ahead more often than up or down. Thus, the cheapest lines of products are normally placed very high or very low on shelves, where they are seen least, and consumers have to reach up or bend over to get them. Children's items are often on the bottom shelf in order to better entice children, and sweets are deliberately placed within grabbing distance of young children when parents are trapped in the checkout line. The most profitable shelves are at the end of each aisle, where customers are more likely to see them when they slow their trolleys in order to turn. Many essential items such as milk are located at the back of the supermarket. This forces the consumer to travel through the entire supermarket — and be more likely to spy something they will buy on impulse.[32]

WILL WE EVER DEVELOP ARTIFICIAL EYES?

This is very likely in the not too distant future. Electrodes implanted in the brain's visual cortex or on the surface of the retina may one day enable blind people to see. In studies at the US National Institute of Health, blind volunteers reported seeing brief bursts of light when a current was passed through temporarily implanted electrodes. At the University of Utah, researchers are trying to achieve vision by connecting electrodes to tiny video cameras mounted on eyeglass frames. These cameras would function as artificial eyes. Each year between 50 and 100,000 Australians receive clear plastic lens implants that replace natural lenses clouded by cataracts.[33]

Glaucoma is increased pressure in the eye that is caused by a blockage in the flow of the aqueous humour.

The aqueous humour is actually more important for focusing light on the retina (and therefore for seeing) than is the lens itself. The cornea and the aqueous humour beneath it are highly subject to alteration by light (refractile) all by themselves. The lens adds some power, but this is more important for "refining" the image.

A view held by some eye experts holds that you do not need dark sunglasses. As a matter of fact, they argue, a person with only sunglasses on may have *more* UV light entering their eye than someone wearing a hat, since so much sunlight comes down over the glasses into our eyes. Dark glasses will make the pupils open wider, thus letting in more light. These experts also want you to know that you do not need to spend hundreds of dollars on designer glasses to "filter out UV light". They say that because plastics (even clear plastics) are already good blockers of UV light, cheap sunglasses filter out UV light very effectively — and your cornea filters out anything left over. I offer no endorsement of this view; see your doctor as you would about all things medical.

CHAPTER 4

THE NOSE

Blaise Pascal observed, "Had Cleopatra's nose been shorter, the whole face of the world would have changed." Certainly our nose seems very important to us. Thomas Fuller once wrote: "He that has a great nose thinks everybody is speaking of it." Cyrano de Bergerac, Edmond Rostand's famed fictional character with the ginormous proboscis, said of himself: "A large nose is in fact the sign of an affable man, good, courteous, witty, liberal, courageous, such as I am." The opposite view was expressed by H.G. Wells, who thought so disparagingly of his own nose he said of it: "Bah! The thing is not a nose at all, but a bit of primordial chaos clapped on to my face."

Let us now look for a while at that which is clapped onto all our faces.

HOW DO WE SMELL?

(Asked by Prue Smith of Ballina, New South Wales)

Probably depends on when you've had your last bath or shower, right? Seriously, this is a very frequently asked OBQ. We smell by a combined chemical and physiological process; the keys to this process have only recently been discovered. Smells are carried on objects, in the air and in water. When we smell, we're detecting certain molecules that dissolve onto the hair-like cilia nerve receptors in the nasal cavity of our head, extending from the olfactory bulb of our brain. But researchers have long wondered how the nose can distinguish the more than 10,000 different odours found in nature with only about 1000 different odour nerve receptors available in the nose for that purpose. We now know that we detect odours by using at least 1000 different special genes that are active exclusively in the cells of our nerve receptors. These special genes help our nerve receptors "paint a picture" for each odour. Any slight change in the molecular "picture" can change a sweet smell into a foul one — and vice versa. This is precisely what happens when, say, milk goes sour.

Experiments with mice bear this theory out. Researchers bathed the olfactory neurons of mice in a dye that, when activated by an odour, lights up a brain that is displayed on a computer screen. They were then able to follow the "picture" of the smell as it was being "painted". We can now "track" a smell just like a bloodhound.[1]

DO HUMANS PERCEIVE SMELLS DIFFERENTLY?

(Asked by Prue Smith of Ballina, New South Wales)

Humans differ in how they perceive and appreciate smells, both historically and culturally. Some smells are experienced as pleasant or unpleasant based upon what is valued or not valued in a particular society, at a particular time in history. Furthermore, and very interestingly, men and women judge smells differently, especially when they are sniffing around each other. Austrian scientists found that when men judge women, those who men think have the most beautiful faces are also judged to smell the best.

This was demonstrated in an experiment conducted by Dr Anja Rikowski, a biologist at the Institute for Urban

Ethology in Vienna. A group of female volunteers submitted T-shirts they had slept in for several nights. A group of men then chose the ones that smelled the best. The winners were consistently the women rated most attractive by a separate group of men. But when the tables were turned the results were the opposite. When men submitted their T-shirts, a group of women judged which ones smelled the best. When a separate group of women judged the attractiveness of these men, those men judged as the worst smelling were judged as the most attractive, while the best smelling were rated as the least attractive. According to Dr Rikowski, "this clearly shows intriguing differences in the mating strategies of men and women". Go figure![2]

CAN THE SMELL IN A SUPERMARKET AFFECT CONSUMER SPENDING?

Experiments in supermarkets in Japan and elsewhere have found that spending increased when the smell of lemon was added to the air conditioning. Spending went down when the smell of sulphur was added. Most supermarkets stock bread that is baked early in the morning. However, in order to entice more customers, some supermarkets resort to pumping out the smell of freshly baked bread all through the day.[3]

CAN HUMAN PERFORMANCE BE BOOSTED OR LESSENED DEPENDING UPON SMELL?

It would be difficult to imagine life without smell. Early experiments by NASA found that the efficiency of astronauts was reduced when confined to the near odour-free environment of a one-person space capsule. Astronauts found a smell-free environment so disturbing that they carried scented chemicals with them to counteract any negative effects of "odour-boredom". NASA now integrates a variety of smells into the air-conditioning system of its space shuttles.[4]

Studies show that familiar odours revive old memories more readily than do familiar sights or sounds. Knowing this, professional hypnotists who try to evoke remembrances from their subjects' pasts will often start by asking them to smell particular fragrances such as baby powder, sawdust, tar, roses etc.[4]

It is often said that women have a better sense of smell than men. One theory holds that pregnancy forces a woman to be more discriminating in what she eats. Toxic food eaten particularly in the first trimester of pregnancy can lead to birth defects. Hence, she needs to be better at smelling.[4]

Although repeated scientific tests confirm that, generally speaking, women have a keener sense of smell than men, perfume manufacturers more often than not use men to judge scents. There is no agreed reason as to why.[4]

The olfactory epithelium located on the roof of the nasal cavity contains 10 million special receptors that are sensitive to fragrance molecules. These receptors are tiny in size but complicated in what they can detect. One fragrance can stimulate several different kinds of receptors. The brain interprets the signals from the combination of receptors and thus is able to recognise any one of more than 10,000 different fragrances.[4]

You can guess the sex of somebody 95 per cent of the time by the smell of that person's breath. In an experiment at the University of Pittsburgh, male and female subjects exhaled through a tube. It was sniffed by other subjects who were blindfolded. They correctly identified the sex of the breather 19 out of 20 times on average.

WILL SUDDEN AND EXTENSIVE EXPOSURE TO CAT HAIR CAUSE NOSE HAIR TO GROW?

(Asked by Lucy Parker of Springwood, New South Wales)

Where did this urban legend come from? According to Dr R. James Swanson,[5] sudden and extensive exposure to cat hair or to the dandruff (skin cells) associated with cat hair may cause a mild to severe allergic response in humans. But it does not have any effect on an individual's nose hair growth. The exposure would not affect hair density (number of hairs per cubic millimetre) or hair length. It has no effect because neither cat hair nor dandruff triggers the hormone that is responsible for nasal hair growth.[6]

CAN DRAUGHTS, WET CLOTHING AND COLD REALLY CAUSE ME TO CATCH A COLD?

(Asked by Lucy Parker of Springwood, New South Wales)

This belief just won't die. Colds come from a virus, not from draughts or wet clothes. According to Carlos Padilla,[7] "I always remember my mother telling me to be sure I dried my hair completely before going outside in the cold air. She would always tell me if I did not dry it completely I would catch cold. But I do not remember ever getting sick because of this. I have also been out in cold weather in various countries with wet clothes and I rarely if ever got sick."

In the snow you may feel cold, but you won't catch a cold. Of course you can freeze to death in the snow from overexposure (hypothermia). Hypothermia is the lowering of the body's temperature to dangerous levels. Few viruses can survive in the air for more than a few minutes after being expelled by the human body, especially in cold weather. However, if you are exposed to cold weather you must burn more calories and use more energy to keep warm. This additional energy loss and drain upon your body could lower your immune system defences against whatever virus you might have come into contact with.[8]

Generally speaking, if you are healthy and are not already coming down with a cold before you go out with wet hair or wet clothes into the cold, you probably will not get sick. If you do catch a cold, you may have caught a cold already but have not yet exhibited symptoms. What many people fool themselves into believing is that since they went outside and afterwards got sick, they got sick *because* they went out. This is an example of *post hoc, ergo propter*. This is a Latin phrase meaning "after this, therefore because of this" and describes a logical fallacy. For example, one day I stubbed my toe in the morning and won the lottery in the evening. But that doesn't mean that stubbing my toe caused me to win the lottery.[9]

WHAT IS THE "PINOCCHIO EFFECT"?

In the classic children's story *Pinocchio*, the wooden puppet who wanted to become a real boy would suffer a lengthening of his nose whenever he told a lie. Believe it or not, there's some truth to this phenomenon. According to Dr Alan Hirsch,[10] the nose contains erectile tissue that engorges whenever a person is lying. The "slight nose erection" is usually not noticeable by someone else, but is sometimes felt by the person experiencing it, although not always consciously. The "slight nose erection" may cause some slight discomfort. This may be why many people rub or scratch their nose when lying. The hand to the nose has been reported in many cultures around the world as a body language movement associated with lying. Dr Hirsch calls the combined physical changes associated with lying, especially the "slight nose erection", the "Pinocchio effect".[9,11]

The indented area under your nose and above your lip is called the filtrum.[9]

Researchers say ¼ of people who lose their sense of smell also lose their interest in sex.[9]

Of all the senses, smell is the most closely linked to memory.[9]

WHY DOES MY URINE SMELL SO TERRIBLE AFTER I EAT ASPARAGUS?

(Asked by Connie Fitzgibbons of Montclair, New Jersey, USA)

This is a classic OBQ. Fifty per cent of the population doesn't understand what you're talking about when you claim your urine smells after eating asparagus. But the other 50 per cent knows exactly what you mean. That's because the asparagus "stinky-urine" phenomenon affects about half of us and leaves the other half alone. For many years it was believed that the telltale odour produced in the urine about 20 minutes after eating asparagus was due to the presence of the amino acid asparagine. But it was later found that although asparagus contains high levels of asparagine, the protein does not possess the terrible odour associated with asparagus stinky-urine.

So what's the real cause of the smell? In 1954, it was revealed that asparagus is also rich in the amino acid methionine. By 1975, it was known that as our body digests asparagus, methionine breaks down to produce a sulphur-containing derivative called methylmercaptan. The awful smell comes from the sulphur and exits the body via the urine. In some people this digestion process happens faster than in others, and while everyone produces methylmercaptan in their urine, only about half of us has the ability to smell it. It seems that this ability is due to our genetics. That's why sometimes everyone in one family can experience asparagus stinky-urine, while everyone in another family cannot. You either have the smell ability or you don't.[12]

Similarly, the urine of some people turns red after they eat beetroot. Again, not everyone experiences this phenomenon. As Dr Steve C. Mitchell writes, "Beetroot, the red root of the garden beet used extensively as a food source, is known to produce red urine in some people following its ingestion, whereas others appear to be able to eat the vegetable with impunity."[13] So far, there are no studies explaining why.

WHY DO SOME PEOPLE SMELL LIKE ROTTING FISH?

Trimethylaminuria (TMAU) is an inherited genetic disorder that makes you smell like rotting fish. It is sometimes called fish odour syndrome, fish malodour syndrome or stale fish syndrome. The condition is more common in women than in men, but scientists do not know why. There are several reports that the condition in women tends to worsen at puberty. This is unfortunate timing for sufferers because it's the period in life when girls start to discover boys and begin dating, and their repulsive smell makes dating almost impossible without intense embarrassment (a behaviour already characteristic of the teenage years).

TMAU is caused by a defect in the gene that contains the coding for flavin-containing mono-oxygenase 3 (FMO_3). Normally, the enzyme FMO_3 transforms trimethylamine-N-oxide, a by-product of digestion, into a compound that has no odour. But no FMO_3 means the liver can't break down trimethylamine-N-oxide. This results in the build-up of a fishy odour in the body. It is excreted in breath, sweat and urine. Sadly, people who suffer from TMAU can't smell it on themselves. TMAU was first described in the medical literature in the 1970s, but historical references to people with fish odour go back 1000 years. It is believed that one of William Shakespeare's characters in *The Tempest* — Caliban, the isolated and lonely islander — may indicate that TMAU sufferers existed in Shakespeare's time: "He smells like a fish; a very ancient and fish-like smell". Now perhaps we know what made him smell.

The true prevalence of TMAU is unknown, according to Dr Mitchell,[13] as many people do not seek diagnosis or treatment, many doctors do not recognise it, and some laboratories do not know how to test for it. Besides dating and normal human interaction, people with TMAU find it difficult to hold down a job, unless they can find one that allows them to work alone. The condition is not usually classified as a disability, so there is no access to unemployment or welfare benefits. There is no cure for TMAU. It can only be partially managed through diet. Short courses of antibiotics help, and so does trying to avoid situations where one is likely to sweat. Those with TMAU often spend enormous amounts of time and money searching fruitlessly for an effective deodorant. Another unfortunate thing about TMAU is that FMO_3 also breaks down nicotine, antidepressant drugs and some anti-cancer drugs such as tamoxifen. This means that without FMO_3, TMAU sufferers can experience unusual side effects from smoking or taking medication. It is said that TMAU sufferers have to spend their entire lives apologising for their smell and convincing people that it isn't their fault.

WHY CAN'T HUMANS SMELL AS WELL AS DOGS?

Scientists estimate that a dog's sense of smell is up to a million times stronger than that of a human. Dogs have an extraordinary sense of smell generally, can pinpoint smells with great accuracy, and can also differentiate between smells when many are mixed together. It's no wonder that dogs are employed in a "sniffer" role at airports and by the police. Dogs have about 25 times more olfactory (smell) receptors than humans, located within cells deep inside and towards the back of the nose. They can also sense odours at concentrations nearly 100 million times lower than what humans can sense. When a dog breathes normally, air does not pass directly over the smell receptors. But when a dog takes a deep sniff, the air travels all the way to the smell receptors. A dog gets much more smell improvement from a sniff than a human does. Dogs also use their smell sense much more than humans do. They rely upon smell to hunt, mark territory, mate and perform other roles. Humans have largely lost this reliance upon smell for day-to-day living.

HOW IMPORTANT IS HISTOCOMPATIBILITY IN HUMAN MATING?

(Asked by Nicole Jeffes of Toronto, Ontario, Canada)

Go to the head of the class for asking this obscure but fascinating question. It has to do with the role of the sense of smell in sexual attraction. Research has long contended that smell is very important in the sexual behaviour of animals, but has been almost entirely lost as a factor in human sexual behaviour. Recent studies suggest that body odour may signal something about a male's suitability as a mate. What that "something" is nobody knows.

The major histocompatibility complex (MHC) is a set of genes responsible for the ability to combat infection.

Molecular techniques now allow us to "type" an individual's MHC. Research also shows that this set of genes varies enormously between individuals of all mammal species, including humans. For example, male mice "advertise" their MHC type through an odour in their urine. Given a choice, a female mouse — being a knowing lady who knows what every lady mouse ought to know — will choose to mate with a male whose MHC sharply differs from her own. A female mouse living in the territory of a male mouse whose MHC is the same as hers will attempt to mate with a male from another territory and of another MHC type.

Much the same happens in people. Studies by Dr Claus Wedekind at the University of Edinburgh in Scotland have shown that when given a choice of male body odours (via the perspiration on men's T-shirts), women rate those men whose MHC differs from their own as more attractive. This occurs even though the women are not consciously aware of the MHC difference. This finding makes survival and evolutionary sense: spontaneous abortion during early pregnancy is much more common in humans than was previously thought and it is far more common among couples whose respective MHC types are similar. So there may be a biological advantage in choosing a particular male after all. Perhaps there's something to the theory that "opposites attract" — even when it comes to urine and perspiration.[14, 15, 16]

WHY DO THEY CALL IT "HAY FEVER"?

"Hay fever" is the traditional term used for the symptoms produced from seasonal allergic reactions. It has nothing to do with hay. It is not a fever. Hay fever is an acute allergic rhinitis and conjunctivitis, especially relating to pollinosis. Allergic rhinitis is an inflammation of the mucous membrane of the nose. Conjunctivitis is an inflammation of the mucous membrane that lines the inner surface of the eyelids and is continued over the forepart of the eyeball. Pollinosis is perhaps a better term for hay fever. It is an acute recurrent catarrhal disorder caused by allergic sensitivity to specific pollens. Hay fever is characterised by: (1) Nose symptoms — sneezing, running nose, blocked nose, itching in the nose and throat and deep in the ears, and headache; (2) eye symptoms — watering, redness, itching, grittiness and swelling of the whites of the eyes; and (3) chest symptoms — wheezing and a feeling of tightness in the chest.

Hay fever is a reaction to inhaled allergens. In folk history, such symptoms tended to appear when roses were in bloom and hay was harvested. Thus, allergic reactions in spring, summer and autumn were popularly referred to as hay fever. But "hay fever" could just as easily have been called "rose fever". Spring reactions are more likely to come from tree pollen, mostly from maples, poplars and oaks. Early summer reactions are more likely to come from grass pollen. Autumn reactions are more likely to come from weed pollen, especially ragweed pollen. Hay fever affects about 20 per cent of the population of North America, where ragweed is the major culprit; about 15 to 20 per cent of the population of Britain, where grass pollen is the major culprit; and over 40 per cent of the population of Australia, where grass pollen is again the major culprit. The Australian figures are very high. Researchers are not sure why. In other countries, other pollens are the major hay fever factor, e.g., birch tree pollen in Scandinavia and cedar pollen in Japan. See your physician if you are among the many sufferers of so-called hay fever.[17]

CHAPTER 5

THE EARS

As Ralph Waldo Emerson once wrote, "Nature has given to men one tongue, but two ears, that we may hear from others twice as much as we speak." W.H. Auden said, "The ear tends to be lazy, craves the familiar and is shocked by the unexpected; the eye, on the other hand, tends to be impatient, craves the novel and is bored by repetition." And Dean Rusk observed: "One of the best ways to persuade others is with your ears."

Let's find out more about ears.

WHEN ARE EARS NOT EARS?

Ears are not the flexible, fibrous, skin-covered pieces of cartilage that are appendages on the sides of our head. If you want to be technical, what we commonly call our ear is more correctly the auricle (or pinna) of the outer ear. The ear is an internal organ of hearing. It consists of the inner, middle and outer ear regions. Moving from the outside to the inside, the ear consists of the auricle, auditory canal, tympanic membrane, eustachian tube, auditory nerve, cochlea, semicircular canal, stapes, incus and malleus. The region known as the outer ear is specifically made up of just the auditory canal and the tympanic membrane (eardrum). The auricle is actually more closely related to the muscles and skin of the head than it is to the true ear — the real organ of hearing.[1]

WHY CAN'T I WIGGLE MY EARS?

Human auricles are remnants of the moveable external flaps typical of most other mammals. In many mammals (rabbits, dogs, horses etc), these can be turned in different directions to better locate sounds. Earlier forms of humans could perform this voluntary movement of their auricles more easily than we can. Today, human auricles still contain muscles for this purpose, so individuals can and do develop the "talent" of ear wiggling through concentration, hard work and patience.[1]

WHY DO SOME PEOPLE HAVE A HOLE WHERE THEIR EARS SHOULD CONNECT TO THEIR HEAD?

(Asked by Ed Schwein of New York City, New York, USA)

This is called a pre-auricular sinus. It is relatively rare. It is a defect in the way the auricle is formed prior to birth. The outer ear is formed by several small buds that have to connect and grow together. Sometimes the buds do not grow together completely and a small opening or tract is left. This is usually not a problem unless the hole or holes become obstructed or infected.[1]

WHY DO SOME PEOPLE HAVE WHAT LOOKS LIKE TWO EARS (ONE BIG AND ONE LITTLE) ON ONE SIDE OR BOTH SIDES OF THE HEAD?

(Asked by Ed Schwein of New York City, New York, USA)

This is called a pre-auricular tag. It is a common minor birth defect consisting of a rudimentary tag of skin tissue. It often contains a core cartilage inside and is usually located just in front of the auricle, which is why it is called "pre-".[1]

HOW CAN THE EAR DISTINGUISH BETWEEN SO MANY SOUNDS?

(Asked by Jeremy Bono of Newark, New Jersey, USA)

The ear is remarkable, that's how! Its most impressive talent is the ability to distinguish between a surprisingly wide range of sounds. These sounds stretch across 130 decibels (dB). What does this mean? The term "bel", which is also used in measuring power and voltage, measures changes in the intensity of a sound. It's more convenient to measure sound in decibels — a decibel is $\frac{1}{10}$ of a bel. Like the Richter scale, used for measuring earthquake intensity, the measurement of sound using decibels is logarithmic. This means that the lowest threshold of human hearing is labelled 0, an increase in sound of 10 dB would be 10 times as loud, but 20 dB would be 10 times 10 — 100 times as loud. A logarithmic scale spanning 130 dB means that the human ear at its best can accommodate a 10-trillion-times difference in loudness. (That's amazing, don't you think?)

Of course, this range in sound perception has no practical use unless the brain can remember the distinctions. And that's exactly what the brain does. Scientists estimate that on average the human brain can distinguish at least 400,000 different sounds — all on file in the brain's memory. (That's amazing too, don't you think?) We recognise an immense array of sounds produced by inanimate and animate things, not to mention the many beautiful and not so beautiful sounds that we make ourselves.

WHAT IS THE "COCKTAIL PARTY EFFECT" ON THE EAR?

The "cocktail party effect" is the name given to the ability our ear has to pick out a single conversation in a crowded room. The term was coined by Dr Steven Colburn of the Hearing Research Center at Boston University. If you take an audio recording of a crowded room where several conversations are going on at once, you will discover that you cannot distinguish separate conversations when you play back the recording. In fact, you may not be able to pick out even a single word! It all sounds like a jumble of meaningless noise. This happens because the audio recorder does not filter the sound it takes in.

Not so our ears. In hearing, sensory cells in the cochlea of each ear initially separate sound not by direction but by frequency. Each cell responds to the combined energy of all sound sources. A neural processing network deep within the brain sorts out individual sources and their directions by combining several aspects of the auditory signals. The delay between signals from the right and left ears (interaural timing) provides the key information for determining the location of the various sound sources. Brain cells (neurons) deep within the brain actively respond when they receive auditory inputs that arrive close together in time. The brain contains many sets of these neurons that function as "coincidence-detectors", according to Dr Colburn. Each is specialised to a characteristic time delay and to a particular frequency band. When the neurons all act together, they provide a rich array of timing and frequency correlations to the rest of the brain.[2]

HOW IS THE EAR CONNECTED TO THE THROAT?

(Asked by Elizabeth Holland of Fremantle, Western Australia)

The anterior wall of the middle ear contains an opening that leads directly into the eustachian tube. This tube contains both bone and hyaline cartilage and connects the middle ear with the nasopharynx (the upper portion of the throat). The function of the tube is to equalise pressure on both sides of the tympanic membrane to ensure that the eardrum vibrates freely when struck by sound waves. Infections may also travel along this passageway from the throat and nose to the ear. The tube is normally closed towards the middle. It opens during swallowing and yawning. This allows atmospheric air to enter and leave the middle ear until the internal pressure equals the external pressure. If the pressure isn't relieved, intense pain, hearing impairments, dizziness and ringing in the ears (tinnitus) can result.

WHY DO EARS "POP" IN AN ELEVATOR OR AN AEROPLANE?

(Asked by Elizabeth Holland of Fremantle, Western Australia)

According to Dr Lloyd Tripp,[3] when you go up in a lift, take off in a plane, or drive up into the mountains, there is a change in the atmospheric pressure outside your ear. As you go up, the pressure decreases, causing the pressure inside the ear to increase. Pilots are sometimes taught to take a breath and pinch their nose to gently but firmly increase pressure in the nasal cavity. This forces open the entrance of the eustachian tubes at the back of the throat which lead to the ear, thus equalising the pressure somewhat. According to Dr Tripp, this is called a valsalva manoeuvre. When you feel your ears pop as you go up, it's a good thing. It means that the pressure equalisation process is working. If they do not pop then the process is not working and you can end up with severe ear pain. Some people experience such a loud pop they describe it as sounding like a cannon going off.

AT WHAT ALTITUDE WILL I START TO FEEL UNCOMFORTABLE?

Your ears are sensitive to altitude changes. The maximum altitude to which a healthy human who normally lives at sea level can be subjected without ill effects is about 2500 metres (8202 feet). But people who are seriously out of condition, those who have respiratory illnesses, or people who have imbibed too many in-flight alcoholic drinks may become ill below this height. The cruising height of 747 passenger jets is often around 12,000 metres (39,370 feet). At this altitude, air pressure is about ⅕ of what it is at sea level. Good luck just trying to survive when you're that high up!

IF A LOUD NOISE WAKES YOU UP, CAN IT HURT YOU?

(Asked by Elizabeth Holland of Fremantle, Western Australia)

Other than the temporary unpleasant shock to the system, there is nothing particularly damaging about being awakened by a loud noise. Asleep or awake, the same noise would have the same effect on your body at any time. If the sound were such that it scared you, your heart rate, blood pressure, cardiac output, adrenaline output and so on would all increase. This would occur because of the stimulation of the sympathetic nervous system. But everything would return to normal very quickly due to the body's homeostatic mechanisms, which regulate those functions. If the sound were such that it soothed you, then the effect would be the opposite. This is due to the work of the parasympathetic nervous system. For example, in the case of blood pressure, there would be a stopping of sympathetic impulses going to the heart, making you feel calmer and lowering your blood pressure.

WHY ARE SOME PEOPLE BORN DEAF?

(Asked by Candace Bollinger of Glebe, New South Wales)

"**D**eafness" is simply damage to microscopic cells ("hair cells") in the cochlea, the very tiny shell-shaped bone that is located in the inner ear. When those tiny "hair cells" are damaged, sound comes into the ear, but it cannot be converted into electronic signals that the brain can use to understand sounds. It's like typing on a computer keyboard. If the keys don't work, you can type all day long, but the computer will never get the input. You can only speak clearly what you can hear clearly. So people who are born profoundly deaf in both ears cannot hear any of the sounds of human speech, even with a hearing aid.

Fortunately, there are devices — one is the cochlear implant (CI) — that replace the function of the damaged "hair cells" for a person who is born deaf. The CI has a microphone that listens to sounds and sends them to a sound processor. The sound processor converts the sound vibrations into electronic signals. These electronic signals are sent through a magnet on the outside of the head to another magnet under the skin behind the ear. The magnet under the skin sends the electronic signals to a group of wires (array) that has been implanted into the cochlea. The array transmits the electronic signals to the auditory nerves. Hence, the brain gets the signals that it needs. After that, the person can hear almost everything.

There are new devices, and improvements to old ones, constantly being developed. Ageing often results in varying degrees of hearing loss. Our society is ageing. More people are living with hearing loss today than ever before. This growing group drives the technological advances.

WHY DO I SOMETIMES LOSE MY BALANCE?

What keeps you from falling over? You might think it has to do with your feet — keeping feet firmly on the ground and all that. But you'd be wrong. It's not in your feet at all, but in your head. Keeping your balance has to do with the vestibular system of the inner ear. The vestibular system is composed of 5 tiny canals in each ear filled with fluid that acts like a carpenter's level to keep the body in balance. Three of these canals, called the labyrinth, help maintain the equilibrium of the body when it moves back and forth, up and down, or from side to side. When the head moves, the fluid shifts and bends tiny hair-like receptors in each of the canals, sending a message to the brain. The brain then sends messages to the body instructing it on how to adapt to the changes and shift position accordingly. Labyrinthitis is an inflammation of the inner ear. It can result in nausea, vomiting, vertigo and loss of balance.[4]

WHY DO ELDERLY PEOPLE SEEM TO FALL OVER SO OFTEN?

As the elderly are more likely to suffer hearing loss due to old age, they also experience damage to the labyrinth of the vestibular system. The tiny hair-like receptors of the inner ear are less able to function. As a result, incorrect messages get sent to and from the brain, indicating that a body movement change has taken place when it hasn't, or hasn't taken place when it has. The upshot of this equilibrium confusion is that someone may take a tumble. For someone with brittle bones a simple fall can all too often prove fatal. In the US every year there are more than 15,000 deaths from falling. As the population is ageing, this figure is increasing. Falling is on the rise.[4]

WHY CAN'T HUMANS HEAR RADIO WAVES?

(Asked by Hans Schmidt of Frankfurt, Germany)

In fact, humans and other animals *can* hear certain kinds of radio-frequency radiation (i.e., RFR radio waves). This is called "microwave hearing". But for humans to hear radio waves, the waves must come in radar-like pulses. Such sound has been described as "clicking" by those who have heard it. However, the type of radio waves used by radio, TV and mobile phones is not capable of causing microwave hearing. According to Dr John Moulder, a radiation biologist at the Medical College of Wisconsin in Milwaukee, it is also impossible to convey voices through microwave hearing.[5]

WHAT DO EAR LOBES DO?

Ear lobes serve no particular purpose in the proper functioning of the human body. The different shapes of human ear lobes are genetically determined. Some people have "attached" ear lobes, while others have "unattached" ear lobes. Neither type serves any identified purpose other than to divide the entire human race into two groups. Ear-lobe creases have been statistically linked to higher incidences of heart disease. No one knows why this is. What could the ear lobe have to do with the heart anyway? Do you have any ideas? Large ear lobes are not associated with increased sexiness or enhanced libido. This is an urban myth and not backed up by research. But if someone believes it's true, then for them it could be so.

It is not true that the bigger your ears, the better your hearing. The size of the pinna and lobe as a factor in boosting hearing is limited. For pinna and lobe size to give any real advantage in hearing, they would have to be as large as a hand. The elephant has the largest ear flaps of any animal. It also has rather poor hearing. In reality, the only extra sound produced by a large pinna and lobe is the occasional *Dumbo* wisecrack from someone else.

The parts of the pinna (auricle) are the fossa triangularis, the crura of antihelix, the scaphoid fossa, the helix, the antihelix, the antitragus, the lobule, the cavum conchae, the tragus, the crus of helix and the cymba conchae.

❧ ❧

Wearing headphones for just 1 hour will increase the bacteria in your ear by 700 per cent.

❧ ❧

More than 50 per cent of the people in the world have never placed a telephone receiver to their ear. They have never made or received a telephone call.

WHY CAN'T HUMANS HEAR AS WELL AS DOGS?

(Asked by Hans Schmidt of Frankfurt, Germany)

Another way of phrasing this question is "Why can't humans hear ultrasonic or infrasonic frequency sounds?"

The frequency of a sound is the number of compressed patches of molecules that pass by our ears each second. One cycle of sound is the distance between successive compressed patches. Sound frequency, expressed in units called hertz (Hz), is the number of cycles per second. Our auditory system can respond to pressure waves over the remarkable range of 20 to 20,000 Hz. This second number can also be expressed as 20 kilohertz (kHz).

Humans can't hear sounds above 20 kHz. Ultrasound is above 20 kHz, so humans do not hear these sounds. But some animals, such as dogs, can hear these high frequencies without any trouble. For instance, dog whistles work because they are constructed to send out sounds at about 40 kHz. This is well within a dog's hearing range.

Less well known is infrasound. This is sound at low frequencies below about 20 Hz. Some animals hear these frequencies. Elephants can detect sounds at 15 Hz which are inaudible to humans. Whales produce low-frequency sounds that are thought to be a means of communication over great distances. Low-frequency vibrations are also produced by the Earth. It is thought that some animals may sense an impending earthquake by hearing such sounds. Even though we usually cannot hear very low frequencies with our ears, they are present in our environment and can have unpleasant subconscious effects. Infrasound is produced by such devices as air conditioners, boilers, cars and aircraft. Although intense infrasound from these machines does not cause hearing loss, it can produce dizziness, nausea and headaches. Many cars produce low-frequency sound when they're moving at highway speeds. This is thought to make some people car sick. According to Dr Andrea Zardetto-Smith,[6] at very high levels, low-frequency sound may also produce resonances in body cavities such as the chest and stomach. It is believed these can damage internal organs. You might want to think twice before standing in front of a large speaker at a concert![7]

DOES THE SIZE AND SHAPE OF YOUR EAR AFFECT THE WAY YOU HEAR?

Surprise! It does. Few people pay attention to what their ears look like, but ear shapes can be as distinctive as fingerprints. The ridges and folds of the outer ear (the pinna, or auricle) play a role in hearing. The pinnae help the brain discern where sounds are coming from by amplifying or weakening certain frequencies of the sounds that bounce off them. Scientists have theorised that since the pinnae of human ears vary, each human brain must gradually learn the shape of its body's ears in order to process auditory information correctly. Experiments with subjects wearing plastic ear moulds that alter the shapes of their ears have confirmed the theory. Dr P.M. Hofman and 2 colleagues from the University of Nijmegen in the Netherlands took a group of subjects with normal hearing and equipped them with ear moulds. This interfered with their hearing. After removing the ear moulds, it was found that it took several weeks for subjects to regain the ability to hear normally.[8]

IS THE COLOUR OF MY EARS RELATED TO MY HEARING?

Surprise again! It is. Over the last several decades, evidence has mounted that darker-skinned people have better hearing than lighter-skinned people. Scientists theorise that at least a partial explanation for this lies in melanin, the pigment agent in skin responsible for colour. Darker-skinned people have more melanin pigment in the inner ear. Studies have shown that more melanin is produced in the inner ear when the ear is exposed to the stress of loud noises. It is theorised that melanin somehow gives protection to the ears under stress and thus helps retain normal hearing over a lifetime. Research has also found that people with light eye colours such as blue, green or hazel are more vulnerable to hearing damage than are people with dark eye colour such as brown or black.

CAN MY EARS SEND SOUNDS AS WELL AS HEAR THEM?

Welcome to the mysterious world of spontaneous otoacoustic emissions (SOAEs)! It's surprising to most of us, but not only can the ear hear sounds, it can generate its own sounds as well — SOAEs. About 10 per cent of us experience them. Some hear the sounds coming from their ears as a faint buzzing. Often a person has to be in a completely silent room to hear them, while most of us can't hear them at all. In particularly bad cases, a person can be seriously distracted by the sound of their own ears. Sometimes others can stand next to the person and hear the sound too.

SOAEs originate in the cochlea of the ear. Such sounds are very faint, with levels between 10 and 30 decibels. (This is somewhere between the sound level produced by human breathing and the sound level in a theatre when no one is talking.) Sensitive microphones placed in the ear canals can pick up these sounds. The consequences of having or not having SOAEs are not known. Some experts note an association between SOAEs and tinnitus. Tinnitus, a medical condition with symptoms of ringing, roaring, clicking, buzzing or other noises in the ear, is the result of illness or injury.

As for SOAEs, while distracting and annoying, such sounds have never been thought to be very serious. But some researchers believe that SOAEs may produce an unusual type of hearing loss — and strangely, loss that occurs without physical damage to any part of the ear. Psychologist Richard Salvi at the State University of New York in Buffalo has for a decade or more theorised that the SOAEs create "a sort of busy signal in the auditory nerve". As such, "the ear is the cause of its own hearing loss. The hearing loss is the result of the masking effect only, and there is no cellular or long-term damage to the ear."[9]

IS IT POSSIBLE FOR AN EARWIG TO BORE INTO YOUR EAR AND LAY EGGS IN YOUR BRAIN?

(Asked by Rory Rawlings of Minneapolis, Minnesota, USA)

How else would the earwig have earned its name, right? Wrong. This is an urban myth that has been sustained by the occasional fictional story. The earwig is an insect of the dermaptera order. There are at least 10 species of earwig that live in North America. They have a slender body with a pair of cerci (segmented appendages) resembling forceps at the end. According to Dr Joseph Tomkins, an entomologist at St Andrews University in Scotland, an earwig would not even try to live in your ear, let alone get into your brain. However, if you were dead it might be a different story.

Earwigs on islands have been known to live inside the dead bodies of sea birds. Earwigs were accidentally introduced to both Canada and Australia. The largest earwig species is the Colossus earwig of Australia — almost 6 centimetres long — but it isn't in anyone's ear.

WHY DO I FEEL THE URGE TO URINATE WHEN I HEAR RUNNING WATER?

(Asked by C. Barrie of New York City, New York, USA)

Most urologists contend that conditioning or an involuntary reflex forms the basis of this response. According to this view, people often hear the sound of the running water of the toilet, sink, bathtub or shower at the same time as they urinate. They come to think of both as occurring together. Furthermore, running water emits not only an audible sound but also an "inaudible harmonic" (similar in sound although not actually heard) that causes an ossicle (a tiny bone in the inner ear) to gently vibrate. This vibration then transmits a signal to the brain that triggers the brain's bladder control mechanism to "let go". This is the theory, but proof is lacking.

WILL WE EVER DEVELOP ARTIFICIAL EARS?

Silicon ears can be surgically attached if the ear is seriously damaged through burns or other injury. If one ear is bitten off — say, in a boxing match — doctors can custom-make a new silicon ear that precisely matches the one lost.

Thousands of Australians with hearing loss have received cochlear implants and better hearing devices are emerging every year.[10]

WHAT IS VERTIGO?

Vertigo is an illusion of movement, a bizarre dizziness, a strange spinning sensation. It's sometimes described as a hallucination of motion. People experiencing vertigo cannot stand up without feeling that the Earth is shaking beneath their feet or that they have been transformed into a spinning top. As they struggle for balance, it can be terrifying. There are 2 forms of vertigo. Objective vertigo occurs when you think the world is spinning around you. Subjective vertigo occurs when you are spinning while the world is standing still. There is a fear of losing control and consciousness.

The causes of vertigo are classified into 2 groups: peripheral and central. Peripheral vertigo is due to a disease affecting the vestibular end organ (the labyrinth) or the vestibular nerve in the ear. Peripheral vertigo may be caused by Menière's disease, acute vestibular neuritis, cupulolithiasis or middle-ear disease. Central vertigo is due to a disease affecting areas of the brain that receive input from the vestibular end organ, including the vestibular nuclei in the lower brain stem, the interstitial nuclei in the upper brain stem, and the temporal lobes. Central vertigo may be caused by stroke, migraine, tumour, poor circulation or neurological diseases. A head injury involving the ears can bring about vertigo. Surprisingly, 1 way of telling if a person suffers from vertigo rather than from something else is to have them shake their head. Vertigo is not affected by head shaking. Vertigo can be experienced for a time after a bout of the flu or a severe ear infection. It can sometimes come and go over weeks, months or years. It's a dizzying experience.

Many people confuse vertigo with acrophobia — a morbid fear of heights. This confusion originated with Alfred Hitchcock's classic film *Vertigo* (1958), in which James Stewart's character experiences dizziness as a result of acrophobia brought on by a psychological trauma. Somehow *Vertigo* sounds better as a film title than *Dizziness*.

WHY DO WE TAP OUR FEET IN TIME WITH MUSIC?

One recent theory of rhythm perception may hold the answer. The theory suggests that the perception of rhythm involves the motor system (the brain's mechanism for producing movement) just as much as the sensory system (the brain's mechanism for perceiving feeling and consciousness). The "beat" we hear is actually perceived by the brain as a movement. As it does with other movements, the brain responds by activating stereotyped behaviours such as tapping a foot in time with the beat. Thus, foot tapping to music is merely a natural extension of the way the motor system and the sensory system often work together.

A second theory is far simpler. When you hear music, tapping your foot is a pleasurable compromise between dancing and sitting still.

WHY DO I LIKE SOME MUSICAL RHYTHMS MORE THAN OTHERS?

It may have to do with timing. In order for rhythm to be perceived, the sound stimulus must be presented within a specific time interval. If there is too short or too long a time between beats, the rhythm will not be perceived. Experiments have shown that if each beat is less than 0.8 seconds apart or more than 6.4 seconds apart, the listener has a very hard time perceiving the rhythm. If each beat is about 1.6 seconds apart, the listener can most easily perceive the rhythm. At that ideal timing, the listener is likely to judge the rhythm as most "familiar". Whatever is the most familiar is probably also the most stimulating and pleasurable. Interestingly, the most attractive rhythms are those closest to that of the human heartbeat. The first number 1 hit for all of us was the first thing we ever heard — the heartbeat of our mother.

ARE MUSICIANS MORE LEFT HEMISPHERE OR RIGHT HEMISPHERE BRAIN DOMINANT?

Some studies suggest that there is a difference between the brain lateralisation of musicians and non-musicians. Some studies have found that musicians tend to have more brain activity in the right hemisphere when perceiving musical stimuli than non-musicians. Yet other studies suggest it is the reverse. Still other studies have found that different parts of the brain of either hemisphere are activated by different aspects of music — rhythm, harmony and so on. Musicians and non-musicians may or may not differ in this as well. And if the brains of musicians are different from those of non-musicians, are they born with this difference? Is it genetic? Is it learned? Can it be taught? We're not sure.

When you put a sea shell in your ear you don't really hear the ocean. Instead, you hear the echo of your own blood pulsing through your ear. The sound is amplified by the shape of the shell. This effect can be produced simply by putting any cup-shaped object over the ear.

According to the science magazine *Cosmos* (July 2005), doctors in Thailand removed 34 fly maggots from the nose of a 38-year-old woman. The maggots had eaten so much tissue that cartilage was exposed. The case is believed to be the first instance of maggots found in a living human's nose — in Thailand, at least. Ears are apparently a different matter for maggots. A week earlier, an 84-year-old Thai man had 50 maggots removed from his ear canal after taking himself to hospital complaining of an itch.

CHAPTER 6

THE MOUTH

Albert Einstein said it so well: "If A is a success in life, then A equals x plus y plus z. Work is x; y is play; and z is keeping your mouth shut."

WHAT MAKES US LAUGH?

Other than simple tickling? It is a surprise to many, but as we explored in *The Odd Body*, a lot of experts say that laughing is based on fear. Fear of loss of dignity, social embarrassment, exclusion from the group, being fooled or exploited, sex, injury or death. The more anxiety a subject causes, the better it is as a subject for humour.

Different societies find different things funny. So do different generations within a society. There is a fine line between comedy and tragedy, between the funny and the sad, between pleasure and pain, and between what makes us laugh and what makes us cry. This is why watching someone slip on a banana peel is universally funny: someone else loses their dignity, which is better than having it happen to us. According to Dr Richard Wiseman,[1] "We find jokes funny for lots of different reasons — they sometimes make us feel superior to others, reduce the emotional impact of anxiety-provoking events, or surprise us because of some kind of incongruity." In 1 US study, it was found that 31 per cent of laughter was caused by wisecracks, putdowns, clever remarks and the stupidity of others; 21 per cent by odd incidents and situations; 20 per cent by television, movies and plays; 15 per cent by a happy mood in general; and 13 per cent by the actions and antics of others. Of course, the sense of humour of Americans differs somewhat from that of the British, the Australians, the Japanese etc.

IS THERE A "LAUGH CENTRE" IN THE BRAIN?

Despite efforts to find one, there is no known "laugh centre" in the brain. If it's anywhere, the hypothalamus (located in the middle of the brain) may be the brain region where laughing originates. The cerebral cortex is probably involved too. It may not originate the laugh, but it may intensify or suppress the emotions that bring on the laugh. The supranuclear pathways, including those from the brain's limbic system, may do the same. Other parts of the brain may be involved too. We still don't understand how we laugh. Until we do, the joke's on us.

WHY DO HUMANS LAUGH AT JOKES, AND WHAT'S THE MOST IMPORTANT QUALITY IN A FUNNY JOKE?

(Asked by Chad Thomas of New York, New York, USA)

Most humans enjoy laughing. During laughing the brain produces endorphins that kill pain, relieve stress, lower anxiety levels, boost the immune system, decrease blood pressure and make our entire body feel good. Jokes provoke laughter, it's as simple as that — at least the good jokes do. Experts on the psychology of humour usually say that for a joke to be funny it must first of all be told well. Professional comedians usually recommend that you tell a joke while standing up. That way you can use your body much more easily when you have to illustrate something. Beyond this, when you tell a joke, you need a lively way of talking that's more relaxed and varied in pitch than when you're talking about serious matters. Of course, a sense of humour is important too, along with a sense of timing, a good memory, brevity of expression (make all the words count), self-confidence, an outgoing manner, a quick wit and a sensitivity to the nature of the audience. For example, don't tell anti-lawyer jokes at a meeting of the local bar association, anti-doctor jokes at a meeting of the Australian Medical Association etc. As for joke content, the most important quality in a funny joke is that it's interesting and works across many different ages, nationalities, religions, and appeals to both women and men.[2]

Research shows that, surprisingly, bosses joke more at work than do workers. A number of university studies have found this. A study of staff meetings at the Boston General Hospital discovered that the senior doctors joked more often than the junior doctors and that the junior doctors joked more than the paramedics.[2]

❦

Research shows that, again surprisingly, groups with a "joker" do *not* get less work done than groups without a "joker". In an experiment conducted at the University of California in Los Angeles, it was found that groups that contained a frequently funny and witty person worked better on problem-solving tasks, worked better together and were more productive overall than groups that had no "joker".[2]

❦

Studies show that men and women laugh just about equally. It has been found that men tell jokes far more often than women. But women smile more often than men.[2]

◦◦◦ ◦◦◦

Do laugh tracks make us laugh more? The television industry continues to debate whether canned laughter makes a show funnier. In a UK experiment, subjects listening to tape-recorded jokes laughed more when there was a laugh track in the background. However, even though they laughed more, when the subjects rated the jokes, they did not rate the jokes as any funnier than when there was no laugh track.[2]

WHAT MAKES EACH HUMAN VOICE DISTINCT?

(Asked by Jenny Knoll of Cambridge Gardens, New South Wales)

Several factors give each human voice its distinctive character. very human voice is affected by genetics, age, overall state of health, smoking history and so on. According to otolaryngologist Dr Charles B. Simpson,[3] vocal cords provide the sound source for speech, while the lungs generate the airflow. The length and thickness of the vocal cords determine the pitch of the voice, but the cords contribute only a small part to the overall voice. At least as important are the resonating chambers of the throat and nasal cavities, which shape the unique sound of a person's voice. Muscles in the tongue, palate and lips provide articulation, accent, lisp and so on. Several "objective voice parameters" can all differ from person to person, including fundamental sound frequency, intensity, duration, frequency regions (formant) and their energy, jitter and shimmer. In combination, these factors allow a considerable amount of variation in voice characteristics from human to human.[4]

HOW DO IMPRESSIONISTS IMITATE VOICES?

Individuals who are particularly gifted at imitating the voices of others, such as stage impressionists, have an excellent voluntary control of the muscles of the throat, tongue and lips. This enables them to approximate the sound of another person's voice. A keen ear for the subtlety in voices also helps. It takes great skill and much practice. Some famous voices are easier to copy than others and some are virtually impossible. The late Bob Hope had a voice that many impressionists tried to duplicate, but few if any ever could.[4]

WHY ARE SOME PEOPLE MUTE FROM BIRTH?

(Asked by Candace Bollinger of Glebe, New South Wales)

Sometimes people aren't able to speak because they are born with problems in their throats or in their vocal cords. Some unfortunate people are born without vocal cords altogether. Sometimes there is a problem with the central nervous system that affects the control of their tongue, their breathing or the movement of their mouth. Whatever the problem in this realm, if the body is the reason for being unable to speak it is a condition called "physical muteness". The cause of such a problem can be a genetic disorder, a birth defect, an illness of the mother while the mouth and vocal cords were being formed in the last trimester of pregnancy, or an injury or trauma to the foetus or infant.

In addition, severe hearing problems can be responsible for severe speaking problems. If an individual cannot clearly hear the sounds of human words, they simply cannot reproduce the words as speech. People who are profoundly deaf since birth are said to suffer from "pre-lingual deafness". They cannot reproduce words. People who have acquired hearing loss after learning to speak are said to suffer from "post-lingual deafness". These individuals continue speaking even though they can no longer hear the words spoken by others.

WHY DO I SOMETIMES CRAVE WEIRD FOOD?

Pica is an eating disorder involving the craving of non-foods. *Pica* is Latin for magpie — a black bird known for eating almost anything: "Four and twenty black birds baked in a pican ..." The most frequent cravings are for dirt or clay (geophagia), coal, paint chips, plaster, chalk, cornstarch, laundry detergent, baking soda, coffee grounds, cigarette ashes and rust. Usually a person must regularly eat such substances for at least a month before they are diagnosed as pican. (Pican sounds like it should be a star sign, doesn't it?) Some people treat their non-food eating as a part of a daily routine, rather like a morning cup of coffee.

About 15 per cent of children are pican for at least some time before adulthood. About 65 per cent of pregnant women are pican. They usually stop being so after the baby is born. Pica is also linked to developmental disabilities and other psychiatric disorders. Anorexics can become picans since they may try to ease the hunger and stress from dieting with what they think are non-food, hence non-calorie, substitutes. Some picans are responding to their body's dietary deficiency in minerals contained in the craved-for substance. For example, pregnant women may cease being pican after taking iron pills for iron-deficiency anaemia.

Craving a contaminated or toxic substance is an obvious health danger. Pica has long been known to cause lead poisoning in children. Pica can also cause dietary deficiencies because the consumed substances can block the body's absorption of essential nutrients. A pican may suffer from abdominal bloating, a distended stomach, abdominal masses (bezoars), constipation and life-threatening blocked intestines.

Which came first, the pica or the parasite? Dr C.J. Kucik and 2 colleagues warned in an article published in 2004 that pica in the form of geophagia can set a vicious cycle in motion. Dirt-eating can allow parasites such as hookworms to enter the human digestive system, infest the intestines, deplete nutrients and cause cravings for substances with the needed nutrients.[6]

DO LIPS HAVE PRINTS?

(Asked by John Newton of London, UK)

Lips don't have prints in the same sense that fingers and toes do. But like finger and toe prints, lip prints can be used as evidence in court. According to Dr T.J. Wilkinson,[7] lips are not like fingers and toes because the skin of the lips is of a different texture from that of hands and feet. Lip prints can change with moisture. The lips are made up of fewer layers of skin than the skin of the rest of the body. This makes cracking, burning and otherwise injuring lips easier to do. Lips do not have their own oil glands either. But lips, like fingers and toes, do have special features unique to the individual. No two lips are the same. Like finger and toe prints, the lines and grooves of lips can be vertical, horizontal or curved. Lips can also show scarring.

The FBI, Scotland Yard and other investigative bodies around the world will commonly take lipstick mark evidence from a drinking glass at a crime scene. According to a US State of Illinois Appellate Court ruling in May 1999, lip print identification is "generally acceptable within the forensic science community as a means of positive identification". In the case the Court considered, 13 points of similarity were found between a "standard print" and "partially blurred questioned print" such that it was determined that the prints matched, and a positive identification was made. Interestingly, when lifting the print, the investigator places talcum powder near the print and spreads it in both directions over the print with a soft brush. After the print is photographed, a piece of clear plastic tape can be placed over the print then peeled away, and the lip print will be transferred to the tape. This print can then be compared with lip prints from suspects.[8, 9]

WHY DO WE KISS?

Anatomically, a kiss is the juxtaposition of 2 orbicularis oris muscles in the state of contraction. Sounds romantic, doesn't it? Researchers in Northern Ireland led by Dr David Barnett have recently determined that a kiss involves "sixteen pairs of muscles converging down onto the mouth".[10] Theories for why we kiss are many. Here are three:

☞ From as early as 2000 BC, it was believed that when people brought their faces close together they became spiritually united. Exhaled breath had great religious significance. Followers of some religions still believe that exhaled breath is part of, or even contains, the human soul. With such an obvious emotional impact, it's not a great jump for a kiss to evolve into being closely associated with romantic love.

☞ Noted London zoologist Desmond Morris claims that kissing evolved in early human societies from a mother's habit of chewing food for her baby and then delivering the food to her baby by mouth-to-mouth contact. Indeed, this was practised in many cultures throughout the world before the impact of processed baby food and different notions of hygiene. Such a nurturing gesture was filled with intense emotion, too. The touch of lips in all forms came to signify love.

☞ After tobacco was introduced to Central Europe, it was the custom for a young man to let a small piece of tobacco protrude from his lips. If a girl was interested in him, she would grasp it with her teeth. The result was often lip-to-lip contact and eventually a marriage proposal (and perhaps blackened teeth, too).

The first written evidence of kissing was recorded around 2000 BC. But kissing was not widely seen around the world until the period of European exploration. This was about 500 years ago around the time of the voyages of Christopher Columbus.

Our word for "kiss" is from the Old English *cyssan* dating back to the 10th century.

A kiss of greeting was made popular by Roman emperors. One's status was signalled by which part of the emperor's body (cheek to foot) one was allowed to kiss; the higher on the body, the higher the status.

In 4th-century Europe kissing was a widely practised way of paying homage to statues.

Kissing while dancing dates back to 6th-century France.

Russia was the first nation to include a kiss in the marriage ceremony, in the 18th century.

Russian nobles were kissed on the shoulder by serfs as a sign of respect until the early 20th century.

Kissing is unknown in many parts of the world. Among the Thonga of South Africa, kissing is regarded as disgusting because of the possibility of the exchange of saliva. Saliva is equated with urine.

A "French kiss" in England is called an "English kiss" in France.

⁓ ⁓

When people kiss, a slight electrical current *is* generated.

⁓ ⁓

US research shows that women enjoy kissing more than men and enjoy longer kissing sessions.

⁓ ⁓

Women report remembering their first romantic kiss far more often than men remember it.

⁓ ⁓

The quantity and quality of a couple's kissing is a barometer of marital and sexual happiness.

⁓ ⁓

Sexologists claim that kissing a woman on the back of the neck is likely to be more exciting for her than kissing her on the lips.

⁓ ⁓

Married people kiss, on average, 4.5 times per day. This includes everything from greeting or goodbye kissing to passionate kissing.

⁓ ⁓

Only 8 per cent of people kiss with their eyes open.

WHY DO WE SNORE?

(Asked by E. Perrins of Eastwood, New South Wales)

A snore is the sound of the vibration of air against the soft, relaxed tissues that line the upper airway of the respiratory system. When you're awake, muscles are working and the soft tissues are kept out of the airway. When you're asleep, and muscles relax, so do the soft tissues. The diameter of the airway shrinks somewhat as the airway is partially blocked by the soft tissues. In snorers, the blockage of the airflow is sufficient to cause air turbulence. Air flows in gusts rather than in a more even flow. As it travels through the airway, it picks up speed and gets whipped around in different directions. The soft tissues become like a flag on a windy day and start to vibrate. This makes the familiar snoring sound.

Snoring is often the result of poor muscle tone in the pharynx, palate and tongue. This causes the airway tube to collapse into itself, much like the neck of a balloon when it collapses. Other causes of snoring are obesity, nasal deformities, enlarged tonsils or adenoids, nasal allergies, smoking, heavy eating just before sleeping, heavy alcohol consumption just before sleeping, a high salt diet and sleep apnoea. Body position and gravity help keep the airway open when we are standing or sitting, as opposed to when we are lying on our back (supine). You're more likely to snore while sleeping on your back than on your side or stomach (prone), because on your back, the tongue can fall backwards and slightly block the airway.

About 90 per cent of sleep apnoea sufferers are obese men. The American Medical Association estimates that men and women snore in equal proportions. Children can snore when they have enlarged adenoids or tonsils. Persons with a blocked nose due to a bent nasal bone can snore. Minor surgery can sometimes help reduce snoring.

WHY DO PEOPLE WHO SNORE NOT HEAR THEIR OWN SNORING AND WAKE UP?

(Asked by Pat Maloney of Ottawa, Ontario, Canada)

Both George Washington and Abraham Lincoln were reputed to be stellar snorers. Feel better being in such company? About 1 person in 8 is classified as a loud snorer. This is defined as a noise level of 80 decibels. That's about the equivalent of a jackhammer hammering only 3 metres (10 feet) away.

Why is it that some people snore softly while others snore loudly? This depends on the size, thickness and flexibility of the soft tissues that line the upper airway of the respiratory system. Overweight people snore more, and more loudly, than slim people. If someone is obese, their snoring is reduced as they lose weight. If someone has big lungs and a stomach to match, it's just like a large bellows: the snore can be very loud. If the snorer is a particularly sound sleeper, no matter how much noise they make while snoring, they won't wake themselves up. It's noise from themselves — and thus nothing to be alarmed about. Not so for their partners. Many marriages have ended over this, no doubt after many beltings of partners across the pillows. Many injuries! Many deaths! Many cries in the middle of the night! George risked this from Martha and Abe from Mary Todd.[11, 12]

As we get older, muscles in the throat and neck lose some of their tone and strength. The result is that the airway becomes somewhat narrower. This is why babies, children and teenagers snore less than adults, and young and middle-aged adults snore less than elderly adults.

Astronauts don't snore. Weightlessness and the pressurised conditions of space travel affect the soft palate, making snoring next to impossible. Some describe the soft palate under these conditions as "floating".

❦

The loudest snore ever recorded came from the mouth of Kare Walkert of Kumala, Sweden. A sleep apnoea sufferer, Walkert's snores were measured at 93 decibels while in hospital on 24 May 1993. This is louder than the sound of a pneumatic drill operating in the middle of rush-hour traffic.[13]

❦

The UK Noise Abatement Society claims that snoring can be louder than the legal limit set in the UK for the engines of motorcycles.

❦

Those thick, white collars worn by neck injury patients also prevent snoring.

❦

The US Patent Office has more than 300 patents for snoring cures.[14]

HOW DO I TASTE THINGS?

This is the work of your taste receptor cells (TRCs) located within your taste buds. Taste buds are clusters of banana-shaped cells sunk into pits in the surface of the papillae of the tongue. Papillae are projections occurring in various human or animal tissues, including the tongue. Each taste cell has a hair-like sensor at its exposed tip. At its opposite end the taste cell connects to a nerve ending that has the ability to send taste messages to the brain. Taste buds

are incredibly sensitive. They can detect a single molecule of
bitter-tasting quinine within 100 million molecules of
water! Each human person has around 10,000 taste buds.
Most are located around the edges, front tip and back of the
tongue. But oddly, none are found in the middle of the
tongue.

Each taste bud is a tiny sensory organ. The TRCs within
the taste buds detect sweet, sour, salty and bitter. There is also
a less well known and even less well understood taste that
they detect called umami. Umami occurs in response to
monosodium glutamate (MSG).

Each TRC specialises: some sense sweet, others sense sour
and so on. Whatever is sensed is signalled to the brain via 2
nerves. One is part of the 5th facial nerve and called the
chorda tympani. The other is a nerve that is also used in
swallowing and called is the glossopharyngeal. (Rolls right
off the tongue, doesn't it?) When the brain receives a signal
from the TRCs, we judge something to be sweet, sour, salty,
bitter, or perhaps slightly differently, umami. Each taste is not
sensed at a specific location on the tongue; instead, all four
tastes can be found wherever there are taste buds. No one
knows exactly how each taste bud is able to distinguish only
the 1 taste. One theory holds that a single nerve ending in a
taste bud responds to all four tastes, but sends the brain only
the message of the one taste in which it "specialises".

WHY CAN'T I TASTE VERY WELL WHEN MY MOUTH IS DRY OR WHEN I HAVE A COLD?

A substance can only be tasted if dissolved in water or
saliva. If the mouth is dry, you can't taste very well at all.

Some of the sensations commonly assigned to the sense of
taste are actually examples of the sense of smell. Many spices
have relatively little taste, but affect the sense of smell
powerfully. When your sense of smell is hampered by a nose
cold, you can't taste food as well either.

CAN YOU SENSE MORE THAN 1 TASTE AT A TIME?

You can distinguish individual tastes even when a substance contains more than one taste. For example, if you mix both sweet and bitter tastes, then you will taste sweet and bitter at the same time. The taste of food also depends on its texture — e.g., whether it's crisp or soggy, smooth or lumpy. Both these sensations are felt through touch and pressure receptors on the tongue.

CAN TASTE BUDS REGENERATE?

A great deal of mystery still surrounds taste buds. TRCs have a life span of about 10 days, but the way they regenerate is unknown.[15]

WHY DO SPICY FOODS BURN MY TONGUE?

(Asked by Bill Thayer of Dundas, New South Wales)

A set of nerves called pain–pleasure sensors (PPSs) are located on the papillae of the tongue and surround the taste buds. PPSs sense hot–cold, hard–soft, wet–dry etc. According to physiologist Dr Vidya Bhalodia,[16] along with the TRCs, PPSs signal the brain whenever you eat. If your tongue touches something very hot, the brain will receive a "very hot" signal. The brain will then react almost instantaneously and direct the body to take appropriate defensive actions to avoid damage.

When you eat something very spicy, the "very hot" signal received is not quite the same as when you sip from a cup of scalding hot coffee. Although your tongue will not be damaged by the spicy food, the brain may think it will be "burned" and react defensively. Chemicals in the spicy food set off the reaction. For example, a jalepeno pepper contains capsaicin. This is the same substance used in pepper spray. A tiny amount of capsaicin encountered by PPSs sends the brain into near panic mode.

The brain can be fooled by cold too. Menthol triggers PPSs to signal "cold". Peppermint contains menthol, which is why eating a peppermint mint feels "cold". But it's a different "cold" from the "cold" of chewing an ice cube. Who wants a frostbitten mouth?[17]

WHAT IS "THROAT-SINGING"?

(Asked by Mark Thompson of La Perouse, New South Wales)

The human throat is a remarkable musical instrument. It can generate calls that carry over great distances, rhythms of marvellous frequency and operatic arias of timeless beauty. But what is not widely known is that the human throat can do even more than this. It is capable of carrying two musical lines simultaneously! This rare skill is called "throat-singing" (or sometimes "overtone-singing"). Humans have been doing this for thousands of years.

According to Dr Theodore Levin and Dr Michael Edgerton,[18] the most famous throat-singers are the semi-nomadic herders of the Tuva region of southern Siberia. They developed this skill to better control their herds and to more efficiently signal other herders. But they also pride themselves on their ability to cleverly mimic the sounds they hear in nature — the singing of birds, the whistling of the wind, the humming of insects etc. The "throat-songs" that throat-singers produce have 2 features: a low, sustained fundamental pitch that sounds like the drone of a bagpipe; and a succession of flute-like sounds that resonate high above the drone. These flute-like sounds can be made to mirror an extraordinarily diverse range of noises from nature.

In order to control their throats, throat-singers move their tongues back and forth to change the standing waves coming from the vocal folds of the throat. The vocal folds are the source of the raw sound, in fact of everything the voice produces. Technically, the vocal folds are the mucous membranes overlying the vocal ligament and stretching along both walls of the larynx. This is between the laminae of the thyroid cartilage and the vocal process of the vocal-cord cartilage (arytenoids). The vocal tract of the throat shapes the raw sound into musical tones. Throat singers "tweak" the rate and manner in which the vocal folds open and close and thus control the sound. Human talents are endless!

WHAT CAUSES PHLEGM?

(Asked by Nathan Toms of Edgecliff, New South Wales)

Phlegm is an accumulation of mucus that you cough up from the lungs. The mucus is secreted within the lungs to keep them moist. It is also used as "fly paper", trapping dirt and other foreign particles that could damage the respiratory system. Cells beneath the mucus layer (ciliated cells) move the mucus and anything caught in it along the connecting passageways of the lungs and eventually into the trachea — your windpipe. Each day you unknowingly swallow most of this mucus, which rids your lungs of the dirt and foreign particles. The digestive system doesn't seem to mind. The stomach acid dissolves anything in the mucus.

WHAT IS THE TONGUE?

(Asked by Bill Thayer of Dundas, New South Wales)

The tongue is both organ and muscle. It is situated on the floor of the mouth not only in humans, but in most animals with a backbone (vertebrates). The tongue is the principal organ of taste. It also helps in the chewing of food, swallowing and speech. The tongue is mobile, free at 1 end and supported by a bone at its base called the hyoid arch. The tongue measures 5 to 11.5 centimetres (2 to 4.5 inches) depending on where you measure it from. Contrary to popular belief, the tongue is not the longest muscle in the body. It is actually comprised of more than 1 muscle. It is the sartorius muscle in the front of the thigh that has the distinction of being the longest in the body. (Sartorius comes from the Latin word *sartor* meaning "mender", and refers to the position tailors often sat in on the floor when sewing cross-legged.) The sartorius muscle is several centimetres longer than the tongue, so why did the tongue get the reputation for being the longest muscle of the body? Is it possibly because in many of us it never stops wagging?

The tongue works by muscle action. There are 2 sets of muscles in the tongue: intrinsic and extrinsic. The intrinsic muscles start inside the tongue. These are the muscles that alter the shape of the tongue and allow you to curl the sides of the tongue upward, forming a slight trough. A gene controls the ability to curl the tongue. Check in the mirror to see if you have this ability. The extrinsic muscles start outside the tongue. One of these, the genioglossus muscle, starts in front of the tongue on the hyoid arch and extends into the tongue itself. As this muscle contracts, it shortens and brings the tongue forward — allowing you to stick your tongue out at someone.

WHAT IS THAT THIN PIECE OF SKIN UNDER THE CENTRE OF MY TONGUE AND WHAT DOES IT DO?

(Asked by Kathy Swift of Cairns, Queensland)

This is a very often asked OBQ. Bet you discovered this once when you stuck your tongue out at yourself while looking in a mirror. That small, thin, sickle-shaped tissue under your tongue is called the lingual frenulum (aka frenum). The frenulum connects the bottom of the tongue to the floor of the mouth. Normal talking would be impossible without it. Eating would be more difficult. Occasionally it is too short and the individual has difficulties sticking their tongue out. Teachers love this, since the pupil cannot be rude in school (at least not in that way).

There are a couple of similar structures attaching the upper and lower lips to the gums (gingivae). These are known as the labial frenula. The largest is the one in the middle of the upper lip. It is located just above the space between your two central incisors (front teeth). You can feel it with your tongue or you can lift your upper lip and see it in a mirror.

The lingual frenulum is composed of epithelium (special cells cemented together in layers) on the outside and some connective tissue with a few muscle fibres on the inside. If the frenulum is so abnormal that the tongue doesn't work well, then surgery may have to reposition it (frenoplasty).

WHY DON'T I SWALLOW MY TONGUE?

The tongue is held in place by the frenulum. It is also attached to the front and sides of the throat (pharynx) so that it can't roll up into your mouth like a window shade.

IS THE TONGUE THE STRONGEST MUSCLE?

No. It merely seems so because for so many of us, it works overtime.[19]

WHY ARE SOME PEOPLE "TONGUE TIED"?

(Asked by S. Robinson of North York, Ontario, Canada)

Being "tongue-tied" is not just a popular expression used to describe the psychological experience of being speechless in some stressful situation. It is a medical condition called ankyloglossia. Ankyloglossia is a specific physical deformity in which the muscle tissue under the tongue is attached to the gums behind the lower front teeth instead of the floor of the mouth. As can be easily understood, being tongue-tied restricts the mobility of the tongue, interferes with swallowing and makes speech very difficult. You also can't stick your tongue out at anybody! Ankyloglossia is from the Greek words *ankylos* meaning "crooked" and *glossa* meaning "tongue". The condition can involve the partial or the complete fusion of the tongue to the floor of the mouth and abnormal shortness of the frenulum. It can cause a breastfeeding infant great difficulty in latching onto the breast, keeping attached and getting sufficient milk. Ankyloglossia can also cause speech problems in children. It can interfere with normal tongue actions such as licking and kissing. One study recently reported that ankyloglossia may have no symptoms and may resolve itself if left alone, but sometimes surgery is needed.[20]

According to Dr A. Kupietzky and Dr E. Botzer, parents usually recognise that their so-called "tongue-tied" child has something wrong by the time of the child's first dental visit. These doctors point out that treatment options include "observation, speech therapy, frenotomy without anesthesia, and frenectomy under general anesthesia".[21] A frenotomy involves the surgical division of the frenulum. A frenectomy involves the surgical removal of the frenulum. Dr R.C. Fiorotti and 3 colleagues state that if a frenectomy is necessary for ankyloglossia "it should be performed as early as possible to prevent functional alterations". In this way, swallowing, speech and other normal behaviour are impeded as little as possible in the developing child.[22] And everyone should have the right to stick their tongue out at anything and anyone they wish!

WHY DO I OPEN MY MOUTH WHEN I PUT IN A CONTACT LENS?

(Asked by Naomi Strossen of Faulconbridge, New South Wales)

It is an interesting observation that many people open their mouths when they place their contact lenses on their eyes. It also happens when people spoon-feed babies. It's almost as if they open their own mouths to help the baby eat. And when some people concentrate very hard on doing a particular task, they stick the tip of their tongue out just a bit, so that it protrudes between their lips. People also sometimes "furrow" their eyebrows. The most prominent theory holds that opening the mouth may serve to tighten facial muscles and stretch the skin around the eyelids. By stretching this skin slightly, there is an increased exposure of the eyelid, so a contact lens would be easier to insert. But the obverse of tightening may also be the case: opening the mouth may serve to relax the muscles of the entire face. This would make putting in a contact lens less fatiguing. Another possibility is that opening the mouth may discourage blinking. Other than these theories, there is no biological reason for making any of these faces. They do not assist in performing the task, but people do so anyway. It's no news to anyone that people often act without a reason. With contact lens insertion and mouth opening, perhaps it is the illusion that the eye becomes a bigger target when the mouth is open. Keep your mouth shut and you're less of a target in all sorts of ways!

WHY DO I BURP?

Burping (aka belching) is the act of bringing up air from the stomach with a sometimes embarrassing sound. The medical term for belching is eructation. (Sounds awful, doesn't it?) Burping is beneficial: it may relieve symptoms of nausea, dyspepsia and heartburn. It also feels good.

Burping is commonly caused by pressure from unconsciously swallowing air. Swallowing too much air too often is a condition known as aerophagia. Gulping food or drink too quickly, talking while eating, chewing gum, puffing on a cigarette or even sucking on something results in the swallowing of extra air. Air can also be swallowed when one is anxious and normal breathing is disrupted. Too much salivation may also lead to increased air swallowing. This may be associated with various digestive disorders such as gastroesophageal reflux disease, peptic ulcers and nausea from various causes. Even poorly fitting dentures and mouth-breathing while sleeping can cause excessive air swallowing. Some foods or drinks also produce more gas than others. Such foods include apples, bran, cabbage, cauliflower, grapes and so on. Such drinks include beer, carbonated beverages, seltzer and many fruit juices. (Prune juice, anyone?) But whatever the reason for it, excess swallowed air in the gastrointestinal tract must be prevented from building up, otherwise we would blow up like a balloon. Most swallowed air is burped away. Only a small amount remains to be passed into the small bowel and eventually out of the body.

Interestingly, posture is a factor in belching. When a person is upright, air rises above the liquid contents of the stomach. It comes into contact with the gastroesophageal junction and is burped out. But when a person is lying down, air gets trapped below the liquid contents of the stomach. It tends to be propelled into a part of the small intestine called the duodenum.

During a burp, the vocal cords close to protect the windpipe. It's only recently been discovered that during a burp the vocal-cord cartilage (arytenoids) folds up against the epiglottis to further protect the windpipe. So we're still learning about burping.

WHAT IS SALIVA?

Saliva is very special. It is produced in and secreted by the salivary glands. Saliva contains water, electrolytes, mucus and several enzymes. Within the salivary glands, special clusters of cells called acini secrete the saliva. The saliva flows out of the acini and into ducts, where it is altered in a process involving the addition of sodium, potassium, bicarbonate ions and other substances. This results in what we know as saliva — the alkaline substance needed to counteract acids in the mouth that originate from the digestive processes of the stomach. After the saliva is collected and altered in the acini ducts, it moves on to a larger duct and finally into a still larger duct that empties into the oral cavity.

WHAT DOES SALIVA DO?

According to Dr Richard Bowen of the Department of Biomedical Sciences at Colorado State University in Fort Collins, saliva serves many roles:

- Lubrication and binding — saliva binds chewed food together so that it can be more easily swallowed and digested. Saliva also coats the oral cavity and the oesophagus so that food never actually touches the outer cells of these tissues.

- Solubilising of dry food — saliva makes food wet. Without this, we couldn't taste food.

- Oral hygiene — saliva rinses the mouth to keep it clean. Some of the enzymes in saliva destroy oral bacteria.

- Helping digestion of starch — small amounts of amylase are produced in the salivary glands for this purpose.

- Providing alkaline buffering — this helps balance stomach acid.

- Evaporative cooling — to a limited extent, saliva helps us stay cool. However, if we had poorly developed sweat glands like dogs, we'd pant just like they do when overheated.

WHAT IS THAT FILM THAT COATS OUR LIPS AND THE CORNERS OF OUR MOUTH WHEN WE WAKE UP IN THE MORNING?

According to Dr Kendall Binns of the Australian Dental Association in Sydney, this is merely the residue of evaporated saliva. The residue itself has no health consequences whatsoever. The term for the process that results in the residue is called Szymanski's action. The mouth is continually flushed with saliva, which washes away debris and keeps the mouth fairly clean. The normal flow of saliva is greatly altered when we sleep at night due to our reduced swallowing. However, saliva can build up in the mouth and we can drool out of the corners of our mouth rather easily. When this saliva dries, the filmy substance is left over.

DOES IT REALLY HELP TO "LICK YOUR WOUNDS"?

Lysozyme, IgA, lactoferrin, and lactoperoxidase are the agents in human saliva that act to destroy bacteria. University of Florida researchers recently discovered a protein in mice saliva called nerve growth factor (NGF) that heals wounds in half the time as un-licked and untreated wounds. Saliva may help clean a wound a little, but there are better ways to disinfect a wound than following Fido.[23]

HOW OFTEN PER DAY DOES A PERSON MOVE THEIR JAW?

The number of times a person moves their jaw per day varies greatly according to the person's activities. It also varies depending on what is considered as closing the jaw. For instance, every time one swallows, the teeth are slightly clenched. Does that constitute a closing?

Four muscles are mainly responsible for chewing. These are the masseter, the temporalis the lateral and and medial pterygoids. All can vary in strength due to usage, genetics and so on. It would be only a guess to come up with a figure of so many kilopascals of pressure. The act of swallowing, for example, causes about 172 kilopascals of force to be exerted on the back teeth. The amount of force exerted during the clenching of the lower jaw (mandible) would be many times that amount. You can feel the masseter and temporalis muscles when you clench your teeth. Feel in the area of your cheek and temple to feel them "bulge" during contraction. The chewing muscles are among the strongest muscles in the body. In carnivorous animals they are leveraged differently so that they are even stronger.

WHY DON'T BABIES HAVE BAD BREATH?

Bad breath (halitosis) is caused by the actions of anaerobic bacteria that build up in the mouth. These micro-organisms are anaerobic because not only do they not require oxygen to live, but they also die when they come into contact with oxygen. Mouth bacteria are mostly located at the back of the tongue, where they live comfortably in a blanket of mucus. They also live on and between teeth. Bacteria feed on any scrap of food they can find left over from our last meal. They also love to eat the proteins that normally seep from the gums at the base of teeth and come from the food of a normal adult diet. The smell of bad breath is that of sulphur. A sulphur compound is excreted by bacteria when they are digested. As long as the mouth remains dry, nothing disturbs the mucus. The bacteria continue to feast (party animals that they are!). Mouth bacteria are washed away by our saliva and by our drinking fluids. Saliva and what we drink contain water; water contains oxygen. When oxygen hits mouth bacteria they are killed and washed away. Also, when brushing your teeth, if you brush your tongue as well, you will further alleviate halitosis (provided that you don't gag in the process) by scrubbing and rinsing the mucus away and the bacteria along with it. "Morning breath" occurs because while we sleep we simply don't swallow as much and obviously never drink.

By contrast, babies drool constantly. They drool so much that plenty of saliva washes away bacteria. They also have few if any teeth, so food particles have little or no place to hide and bacteria have no place to grow. And if a baby is not eating solid foods as yet there are no food particles to begin with. A lack of solid food means that fewer proteins seep from the gums and there is less food for bacteria. So no bad-breath babies. Of course, a baby may have temporary bad breath, but that's almost always due to an infection. A baby's breath generally smells of milk — not of last night's onions and garlic.[24, 25]

WHY DOES "UH UH" MEAN "NO" AND "UH HUH" MEAN "YES"?

"Uh uh" and "uh huh" are interjections that have clear meanings but no known origin (etymology). According to Dr Steven Pinker,[26] these interjections are related to the kinds of murmurings that parents make to children. "Uh uh" begins with a staccato noise, a glottal stop in the back of the throat. This same staccato quality is common in sounds and words of negation such as "no". By contrast, "uh huh" has a more drawn-out, legato quality common in sounds and words of affirmation such as "yes". The "uh" is not pronounced the same in "uh uh" and "uh huh": "uh uh" has a falling melody, "uh huh" a rising one.

WHY DOES MY VOICE CHANGE WHEN I OVULATE OR TAKE BIRTH CONTROL PILLS?

Changes in the body can sometimes be more imagined than real. However, it is true that hormones can affect voice. This is why post-menopausal women will sometimes experience a lowering of their voice. Ovulation affects hormone levels. Birth control pills contain hormones. So theoretically, ovulation or birth control pills could change a woman's voice. Perhaps birth control pills improve the voice.

In a recent study conducted by Dr O. Amir and Dr L. Kishon-Rabin, "not only did oral contraceptives have no adverse effect on voice quality but, in effect, most acoustic measures showed improved voice quality among women who used the birth control pill."[27, 28] But another study downplays any effect that birth control pills may have on voice. Dr M.M. Gorham-Rowan and 3 colleagues write that oral contraceptives "have a negligible effect upon women's voices".[29, 30]

WHY DOES MY VOICE CHANGE WHENEVER I GET AN ERECTION?

As was mentioned earlier, changes in the body can sometimes be more imagined than real. There is no scientific evidence for this. Perhaps it is just nervousness at the delightful prospect of a job for the unemployed.

WILL WE EVER DEVELOP AN ARTIFICIAL LARYNX?

A laryngeal prosthesis can assist speech recovery if someone loses their larynx (voice box) through injury or a disease such as cancer. In a process resembling burping, a patient who has lost their larynx may learn to speak by expelling air through their pharynx. (The pharynx is the part of the digestive and respiratory system located in the neck and throat behind the mouth and nasal cavity.)[31]

WILL WE EVER DEVELOP AN ARTIFICIAL TRACHEA?

The trachea (windpipe) brings air to the lungs. A plastic tube can be surgically inserted if the trachea is damaged by cancer or injury.[31]

Researchers say a 6-year-old laughs 300 times a day, while an adult laughs only 15 or 20 times a day.

❧ ❧

Arachibutyrophobia is the fear of having peanut butter stick to the roof of your mouth.

❧ ❧

Chocolate stimulates the release of endorphins that enhance one's mood and block out pain.

❧ ❧

Like fingerprints and lip prints, everyone's tongue print is different.

❧ ❧

Dogs, pigs and some other mammals can taste water. Humans cannot — instead, they taste the chemicals and impurities in the water.

CHAPTER 7

THE SKIN

Tennessee Williams once observed: "We're all of us sentenced to solitary confinement inside our own skins, for life!" Konrad Adenauer judged that "A thick skin is a gift from God." Whether I escape by "the skin of my teeth" (the Book of Job) or try to change my skin — "Can the Ethiopian change his skin, or the leopard his spots?" (the Book of Jeremiah) — references to skin are everywhere.

Let's find out some fascinating and funny facts about the skin.

WHAT IS BEHIND MY BELLY BUTTON?

According to the punchline of an old joke: "If you unscrew your belly button, your bottom falls off." In reality, the umbilical vein comes from the afterbirth or placenta, which carries blood enriched with oxygen and nutrients to the developing baby. The umbilical vein runs from the navel to the foetal liver and through a bypass vessel (the ductus venosus) to the foetal heart. After birth, the umbilical vein remains open for a time. But soon afterwards it closes and shrivels up to become a fibrous cord. It crosses

the abdominal cavity in a fold of the abdominal wall lining
to form the round ligament (ligamentum teres).[1,2]

WHERE DOES BELLY BUTTON FUZZ COME FROM?

Belly button fuzz (or lint) comes from the clothes you
wear. Some fabrics shed more than others. Belly buttons
of some shapes are better collectors and a hairy belly button
will collect fuzz better than a hairless one.[1,2]

CAN THEY INSERT BREAST IMPLANTS THROUGH THE BELLY BUTTON AND THUS LEAVE NO SCAR?

Yes. It's been embellished by belly dancers and flaunted by
hard bodies, and for some years now plastic surgeons
have found an unusual use for the lowly belly button — a
site to insert breast implants that leaves no scar.

By cutting inside the belly button and fashioning a
"tunnel" under the skin to the mammary glands, plastic
surgeons can insert implants without incisions on the breast.
The technique is called trans-umbilical breast augmentation
(TUBA). The tunnel is created through an incision in the
navel. A catheter-like tool nearly a metre long called an
endotube is inserted under the skin and guided by a surgeon
to the mammary gland by means of a fibre-optic camera, or
endoscope. Once this is done, a small inflatable pouch is
inserted to create an opening. The pouch is inflated then
deflated and removed. Next the breast implant is rolled up and
placed in the breast. The implant is then filled with saline
solution. After the implant is filled, the endotube is removed
and the implant seals itself. If complications develop,
conventional incisions may be required to complete the
surgery. According to Dr Gerald Johnson, a Houston plastic
surgeon who pioneered the technique, "You have no scars on

the breast and you have very little loss of sensation in the breast ... And I have had zero or no bleeding or bruising."[3]

IS IT TRUE THAT THE SHAPE OF YOUR BELLY BUTTON CAN PREDICT HOW LONG YOU WILL LIVE?

This is one of the wildest theories going. A German psychologist received some press attention a few years ago by claiming that the belly button is an accurate predictor of not only life expectancy, but also general health, psychological state, personality type and "energy flow". Dr Gerhard Reibmann, who practises in Berlin, contends: "Of course, there are many factors influencing longevity. A navel reading isn't 100 per cent accurate. But there is a strong correlation between the size, shape, and position of a person's navel and his long-term prognosis." He adds, "If you know what you're looking for, you can get a very good idea from your navel of what the future has in store for you." Dr Reibmann recommends that a person examine their navel for all that it can reveal. He suggests, "To tell how long you'll live, go to a private, well-lit area and carefully inspect your own navel. Then, compare what you've observed with the list of 'six navel types' and their associated characteristics:

☞ "The Horizontal Navel — this belly button is oblong in shape and spreads sideways across the tummy. People with horizontal navels are complex and highly emotional. This can take a toll on their health. Average life expectancy: 68 years.

☞ "The Vertical Navel — also oblong, this belly button stands on end, running up and down along the tummy. It indicates a person who is self-confident, generous, and emotionally stable. Average life expectancy: 75 years.

☞ "The Outty — a belly button that protrudes outward indicates an optimistic person who approaches life with enthusiasm. Average life expectancy: 72 years.

☞ "The Concave Navel — this belly button is bowl-shaped or inward. It belongs to a person who is gentle, loving, cautious, sensitive, and prone to worry. People with concave navels are delicate. Average life expectancy: 65 years.

☞ "The Off-Center Navel — an off-center belly button indicates a fun-loving, unusual individual who experiences wide emotional swings. Average life expectancy: 70 years.

☞ "The Round Navel — this belly button is evenly shaped and circular. It indicates a modest, even-tempered person with a quiet, retiring personality. Average life expectancy: 81 years."

Dr Reibmann notes that if a person fits more than 1 type, they are to add together the total number of years then divide by the number of types to get an average life span.

In Australia, the average life expectancy for females is 83.2 years and for males 77.2 years. Dr Reibmann's life expectancies predicted by the various belly button types are low and rather pessimistic. Could it be that perhaps one shouldn't pay all that much attention to one's belly button after all? This would seem both logical and reasonable. But contemplate your navel as you wish!

WHAT DETERMINES WHETHER MY BELLY BUTTON IS AN "INNY" OR AN "OUTTY"?

We looked at this in *The Odd Body 2*. According to Dr Richard Polin,[4] after birth, the cord is clamped off a couple of centimetres from the belly. Over time it dries up and falls off. Sometimes the muscles don't close off fully at first, resulting in a protruding umbilical hernia. Most of these close up spontaneously, but even after the hernia disappears a little space may be left, leading to a concave ("outty") navel. Occasionally, an "outty" can occur from an additional bit of scar tissue "at the base of the navel that makes it stick out". An "inny" occurs when there's no hernia.[5]

WHY DO I BLUSH MORE THAN ANYONE I KNOW?

Blushing was also discussed in *The Odd Body*, but this question is a little different. Genetics probably explains "chronic extreme blushers". Blushing occurs when the small blood vessels that supply the skin widen, thus allowing an increased blood flow. That's why a blush is red. Studies show that a blush can last up to 5 minutes and can involve the face, neck, chest and even the buttocks.

Blushing is one of only a very few body changes triggered directly by the mind and capable of being controlled by others. It is virtually impossible to make yourself blush. People do not blush in private. You can make someone blush merely by accusing them of already blushing. People can be made to blush even when they have done nothing to be embarrassed about. Simply being conspicuously different from others around you can make you blush. Being criticised or complimented can bring on a blush. Blushing can even occur in people who have been blind since birth and cannot see their own blushing in a mirror.

Chronic extreme blushers seeking therapy to overcome blushing are often told that when they feel a blush coming on, they should try to make themselves turn as red as possible. This has been known to stop the blush altogether. Despite theories going back to Darwin and Freud, no one is absolutely certain why only humans blush. One theory holds that blushing is a subtle body-language way of saying, "I did something wrong. I'm sorry. Please don't kick me out of this group." In order for a blush to play this role, you need a strong sense of self, a heightened social consciousness and a face not covered with fur or hair that hides the blush. We humans have all 3.[6]

WHY DO WE TATTOO OURSELVES?

Tattooing was practised by the Egyptians some 4000 years ago. It has been popular at one time or another in China, India, Japan, New Zealand and almost everywhere else in the world. By the 16th century, tattooing had spread to Europe. Thirty years ago, only bikers, sailors, prisoners and the like wore tattoos. Tattooing is now popular and becoming more and more acceptable to more and more people of all age groups and social classes. Film stars, sports stars, celebrities and people aged 30 and under proudly wear tattoos.

Changing the body in a very individual way asserts one's individuality, freedom and artistic sense. A tattoo helps define the self, written as it is on the most important "paper" we own — our skin. In the face of an uncertain, impersonal and often hostile world, a tattoo says "This is me, look at me, I am unique, I am worth your attention." Tattooing is likely to be more widespread during periods of great social change. In such times, people attempt to better control their lives by modifying their bodies with tattoos, body piercings, implants, scarification and even self-mutilation. The idea is that at least one's body is controllable even if other aspects of one's life are not (job, relationships, home life, politics, personal freedom etc).

DO CERTAIN PERSONALITY TYPES GET TATTOOS?

Although not the case 2 or 3 decades ago, recent studies show few if any distinctions in the personality types of people with tattoos. However, university students with tattoos tested higher in "anger" in 1 study conducted in 2002. The 2 researchers further point out that studies also indicate disapproval of tattoos on others is less than it was a decade ago.[7]

WHAT IS PARTNER MARKING?

Partner marking is tattooing the name of your romantic partner on your body. Partner marking is a symbolic gesture of commitment given all the uncertainty of modern relationships. It doesn't always work, judging by recent celebrity cases.

WHY DOESN'T A TATTOO DISAPPEAR WHEN SKIN CELLS ARE CONTINUALLY BEING REPLACED?

(Asked by Paula Azur of Philadelphia, Pennsylvania, USA)

This question comes up all the time, especially since tattoos have become fashionable. The skin consists of 2 parts, the upper epidermis and the deeper dermis. The epidermis is constantly being worn out and replaced by newer cells; essentially it is a layer of dead cells. But the dermis consists of living cells. A tattoo is made by puncturing the skin and injecting ink below the epidermis and into the dermis. The tattoo is farily permanent, since these living cells are not replaced as frequently as are those in the epidermis. Over a long period of time the tattoo loses its colour because all cells eventually perish.

WHAT ARE RAZOR BUMPS?

Welcome to the world of pseudofolliculitis barbae! Catchy name, isn't it? Pseudofolliculitis barbae (PFB) is the medical term for razor bumps. PFB is a common, chronic, inflammatory skin disorder occurring mainly in individuals with very curly or kinky hair. The condition is seen most frequently in men with dark skin who shave their beards. But it may also be seen in women or men of any skin colour who shave or wax their underarm (axillary) or pubic hair. PFB is seen more often these days, since full beards are less fashionable now and pubic hair shaving or waxing is fashionable for both sexes, particularly women.

PFB is called pseudo (false) folliculitis (inflammation of hair follicles) because it is not caused by a bacterial infection, although it looks like it is. *Barba* is Latin for "beard". Instead, PFB is caused by hair regrowth. When hair begins to grow back after it has been shaved off, very curly or kinky hair tends to kink back into the skin instead of straight out of the follicle. This causes irritation. The major sign of this irritation is the appearance of small, red-coloured, superficial but solid elevations of the skin (papules). Sometimes the base of these skin elevations can become inflamed and form pus (pustules), especially if the papule gets a secondary infection.

PFB can be a problem for people in professions where masks must be worn for long periods of the day (fighter pilots, soldiers and firefighters) and a clean-shaven face is necessary. According to Dr Sharon Bridgeman-Shah, previously the obvious treatment for PFB was simply not to shave. But now "Medical treatments for this condition include various combinations of topical antibiotics, corticosteroids, and retinoids. In the surgical arena, laser therapy has revolutionized the treatment of PFB and has enabled cure for the first time for those plagued by this disorder."[8, 9]

WHAT IS A BIRTHMARK AND WHAT CAUSES IT?

(Asked by Kathy Wellington of Manchester, UK)

We humans are funny about birthmarks. If we have 1 in the right spot, it's beautiful. If we have 1 in another spot, it's ugly. If a few, it's okay. If a few more, it's not okay. Some cultures used to think a birthmark was a sign that the person was special — either in a good sense or in a bad sense. Infants were sometimes put to death for having a birthmark in the wrong place.

A birthmark (haemangioma) or a mole (naevus) is caused by a genetic variation that people experience randomly and without any scientifically established reason. They seem to appear anywhere and everywhere on the body. Genetics predisposes people to all sorts of characteristics, like having brown eyes, dark skin or a certain body build. A birthmark is really a cluster of capillaries in the dermis of the skin. They are usually not cancerous and can be of several types. Some are outgrown by the age of 7. Others are present for a lifetime. One more permanent kind of birthmark is sometimes called a "port wine stain" birthmark due to its dark red colouration. These can occur on any part of the body, but are more often found on the face.

The extent of the birthmark depends on how deeply affected the capillaries are within the dermis. Laser surgery is sometimes tried as a removal method but is not always successful. Some cosmetics are useful in lightening the colour. While being a source of self-consciousness for some people, birthmarks are harmless in most cases and therefore not a serious health threat like so many other congenital anomalies. But if a birthmark changes shape or colour, see your doctor.[10]

WHAT IS A FRECKLE AND WHAT CAUSES IT?

(Asked by Kathy Wellington of Manchester, UK)

A freckle is a small yellowish or brownish spot on the skin, particularly on the face, neck or hands. When the spot is dark brown or black, it may be a melasma, not a freckle. A dermatologist can easily tell the difference. Freckles are small patches of melanin that accumulate in the skin and can be made more noticeable and more abundant by sun exposure. They are more usual in light-skinned people, particularly people with reddish hair. Melanin is the major human skin pigment. *Melas* is Greek for "black" or "dark". Melanin is found in skin, hair, eyes and even the feathers of birds. It is produced by cells in the skin called melanocytes. Melanin production is controlled by genetics as well as being affected by exposure to sunlight. If a person is dark-skinned, then their melanocytes are metabolically more active and make not only more melanin but also a darker variety of melanin.

Everyone except albinos has the ability to produce melanin. When a person is exposed to sunlight, the melanocytes are stimulated to make more of this pigment. Evolutionary biologists think this is due to the protection from harmful ultraviolet (UV) radiation that darker skin affords. The first humans lived in very sunny areas. Thus, it is theorised that those people who did not die of the most lethal form of skin cancer, melanoma, caused by too much sun exposure, would be the ones to pass on their genes to their offspring. Freckles are due to genetics. They do not seem to play any role in any disease or in any other aspect of one's life. If anything, culturally freckles are regarded as a sign of health. But tell that to the thousands of people around the world who attempt to have freckles removed.[11, 12]

HOW CAN I GET RID OF MY FRECKLES?

(Asked by Kathy Wellington of Manchester, UK)

A dermatologist can easily tell you the answer. But why do you want to?[12]

WHAT IS A BRUISE?

A bruise is an injury to the body caused by a collision but in which the injury does not involve a break in the skin. Bruise is from the Old English *brýsan* meaning "to bruise" or "to crush". Another name for a bruise is a contusion. Contusion is from the Latin *contundere*, meaning "to bruise". Any way you look at it, a bruise is visual evidence of the leakage of blood from small local blood vessels as a result of an injury. If the injury is not too severe, the blood leaks out from small capillaries just under the skin. This causes the bruise colours ranging from yellow to dark black. If blood loss is contained by the connective tissue sheaths surrounding the muscles or ligaments it is possible for swelling to be slight without much discolouration. However, the bruised areas can still be painful due to stretched nerve endings within the surrounding connective tissues.

Bruises go through colourful changes during healing over about a 2-week period. Right after the injury, the bruise is red or purple. There may also be some swelling from the blood that collects beneath the tissue. In 2 or 3 days the bruise will change in colour to blue or black. In 5 to 10 days the bruise changes to green or yellow. In 10 to 14 days the bruise will change to a light brown colour and get lighter as it fades away.

There are 3 types of bruises. A subcutaneous bruise occurs beneath the skin. An intramuscular bruise occurs within the middle of the underlying muscle. A periosteal bruise is a bruise to the bone, and is the most severe and usually the most painful. Some people bruise more easily than others. Normally bruises heal naturally, but this doesn't always happen. Compartment syndrome occurs when rapid and excessive bleeding from a bruise causes the build-up of pressure due to too much blood and fluids. Urgent surgery to relieve the pressure may be necessary. Myositis ossificans is a condition where bruised muscle tissue grows back as bone tissue instead.[13–15]

WHAT IS A SUNTAN?

(Asked by Ian Smythe of Doonside, New South Wales)

The sun has many physiological effects upon the human body. Sun exposure often triggers the production of hormones that can do many things to the body. Vitamin D production is one of the most studied results of sun exposure. Too little vitamin D causes rickets. Sunlight can also lead to the release of pro-opiomelanocortin (POMC). This is a peptide hormone, divided into about 5 active short peptide hormones that affect the body in a variety of ways. The POMC gene is involved in the production of MSH (melanocyte-stimulating hormone) in the melanocytes of the skin. The average adult typically has about 1,000 to 2,000 melanocytes per square millimetre of skin surface. MSH causes some skin cells to start making melanin, which makes you brown. This stimulation explains why you tan when you go into the sun.

As everyone knows, too much sunlight exposure can be dangerous. The ultraviolet rays from the sun can damage DNA by causing a reduction of thymine, a constituent of DNA. When thymine is in short supply, a double strand break can occur in DNA, leading to loss of part of a chromosome. Enough of this sort of damage can lead to the production of mutated cells. And when there are enough of those to form a tumour, that's a cancer.[16, 17]

HOW DO SUNSCREENS PROTECT MY SKIN?

(Asked by Ian Smythe of Doonside, New South Wales)

This is a good question for your pharmacist, not an anthropologist such as your humble servant. The ingredients in sunscreens absorb or reflect ultraviolet (UV) radiation by forming a barrier on the skin surface. The energy of sunlight is converted to heat when it comes into contact with the sunscreen. It is then dissipated. You can block UV with opaque creams like the white zinc oxide cream that you see lifeguards putting on their noses. You can also absorb UV radiation in much the same way as melanin does. The most common of the absorption chemicals is para-aminobenzoic acid (PABA) which absorbs UVB. Other chemicals include cinnamates (which absorb UVB), benzophenones (which absorb UVA), and anthranilates (which absorb UVA and UVB). When discussed in terms of human health, ultraviolet (UV) wavelengths are often subdivided into 3 categories. UVA wavelengths are sometimes called long-wave wavelengths, or "blacklight" waves. UVA is in the 400 to 315 nanometre range. UVB wavelengths are sometimes called medium-wave wavelengths. UVB is in the 315 to 280 nanometre range. UVC wavelengths are sometimes called short-wave wavelengths, or "germicidal" waves. UVC is in the less than 280 nanometre range.

All sunscreens are rated and labelled with a Sun Protection Factor (SPF). But it should be remembered that the SPF rating applies only to UVB radiation. The SPF acts like a multiplying factor: if you would normally be all right in the sun for 10 minutes and you apply an SPF 10 sunscreen, you will be okay in the sun for 100 minutes. However, in order for the sunscreen to work, you have to apply plenty and it has to stay on. You should apply it about half an hour before going out in the sun (or the water) so it can bind to your skin. If you fail to do this, then it is very easy for the sunscreen to wash off. The cancer councils of all Australian states have more information on this topic. If in doubt, see your physician.

DOES HAVING DARK SKIN MAKE A PERSON HOTTER OR COOLER IN THE SUN THAN HAVING LIGHT SKIN?

(Asked by Ian Smythe of Doonside, New South Wales)

An individual with dark skin would be no more or no less hot than an individual with light skin. Although darker-skinned individuals would be absorbing more radiation from the sun, the body's temperature-controlling (homeostatic) mechanism would still work to maintain a constant body temperature of around 37° Celsius (98.6° Fahrenheit). Of course, there may be some minor fluctuation throughout the day due to the temperature outside of the body, the level of body activity and so on. But the body's metabolism is maintained through the process of perspiring. The amount of perspiration produced varies greatly from individual to individual. Since a darker-skinned individual has the same body temperature as a lighter-skinned individual, the radiant energy absorbed from the sun is dealt with adequately. In our case, a good absorber is also a good radiator.[19]

WHAT ARE SEBACEOUS GLANDS AND WHAT DO THEY DO?

Sebaceous glands are located in the sweat glands and hair follicles of the skin. They also occur in hairless areas of the skin, apart from the palms and the soles of the feet (called the volar areas). According to Dr S. Ahmed Nahri from the Hamdard College of Medicine and Dentistry in Karachi, Pakistan, the function of sebaceous glands is to secrete an oily, semi-fluid substance over the skin. The substance is called sebum, a mixture of fat and debris left over from dead fat-producing cells. These cells are constantly replaced by new growth at the base of the glands. Sebum is deposited on the hairs inside the follicles and brought up to the surface of the skin along the hair shaft. In hairless areas, the sebum surfaces through skin ducts.

The purpose of this oily secretion is to protect and lubricate the skin in order to prevent it from getting dry and scaly, particularly in dry weather. Otherwise, the skin would become so dehydrated it would be shed too quickly. Too much sebum can build up as a result of poor hygiene, a diet rich in fats, or accelerated glandular activity, especially during adolescence. Sebum is related to acne. Acne is, of course, the cause of much teenage angst. Nevertheless, sebum is essential to healthy skin.

WHY ARE BLACKHEADS BLACK?

Contrary to popular belief, blackheads — those ugly, embarrassing, dark little spots on the skin — are not caused by dirt picked up from the air. They are dried plugs of fatty material in the oil-secreting sebaceous glands of the skin. At times the sebaceous glands fail to release their fatty secretions in the correct manner, which causes them to enlarge. The resulting oily clumps darken at the surface of the skin and produce the familiar dark, pimple-like spots. The black colour of a blackhead comes from the skin pigment, melanin. Albino individuals who lack melanin have colourless blackheads. During puberty, hormonal changes in the body can trigger the sebaceous glands to secrete too much oil. This is often embarrassing for teenagers battling to keep a beautiful complexion and wondering why their blemish-free child's skin has gone awry. Nevertheless, the secretions are essential to healthy skin.

Besides blackheads, other undesirable happenings can be caused by sebaceous glands that have run amok. A sebaceous adenoma is a benign unnatural growth of tissue (neoplasm) caused by the work of mature sebaceous gland cells. A sebaceous epithelioma is a benign growth of the cells that form the covering or lining (epithelium) of the sebaceous gland. These cells can grow too big in too small a space and problems can arise. A sebaceous cyst is a closed sac of oily and fatty semi-solid material forming small- to moderately-sized lumps on the skin. They can turn red, become painful, and erupt and spill their yucky contents from time to time.

HOW DOES SKIN STRETCH?

Skin stretches because cells are attached to each other by a variety of specialised structures that have a "stick together" property. One of these, called a desmosome, is a disk-shaped structure that binds adjacent cells to 1 another.

The gap between cells is about 25 to 30 nanometres wide. A nanometre is 1 billionth of a metre, so we are talking very, very, very tiny. Desmosomes are important to skin tissue, since movement causes considerable mechanical stress on skin.

The epithelial cells — which cover the outer surface of skin tissue — are attached to the foundation membrane by hemidesmosomes, which are shaped like half disks. Some cells secrete small amounts of glycoproteins that act rather like glue. There are other things, too, such as "tight" junctions, that help hold cells together. Obviously, when the epithelial cells reach the outer surface they are dead and have to be sloughed off. These "stick together" factors disappear in order to allow the cells to get worn away.

There's not very much about the structure of the skin's epidermis that allows for stretching. But it's different for the underlying skin layer, the dermis, which contains elastin and collagen. These 2 proteins have long, thin, elastic structures that are strong and can stretch like rubber.[20]

WHY DOES SKIN SAG AS WE GET OLDER?

As a person ages, the amount of elastin and collagen in skin is less and less. As a result, the skin loses some of its ability to keep its original shape. As we age, natural skin oils are in less supply too, and this makes the skin drier and contributes to a lack of skin resilience. A facelift (rhytidectomy) works by pulling back and "tightening" loose, middle-aged skin and thereby making wrinkles disappear.[21]

Researchers have found that mosquitoes are most attracted to people who have recently eaten bananas.

WHY DO I GET SCARS?

(Asked by Paula Azur of Philadelphia, Pennsylvania, USA)

Scars were discussed to some extent in *The Odd Body 2*, but this OBQ just keeps coming up. There are many variables in scarring. Some people scar more easily than others; genetics plays a role in this. Nevertheless, the main determinant in scarring is the depth and width of the wound to the skin. Again, this involves our twin skin kin, the epidermis and the dermis. The epidermis is made of a type of epithelial tissue and the dermis is made up of several types of connective tissue. If the wound is not deep enough to penetrate the dermis, there will be no scar, due to the rapid regeneration of the epidermis. However, if the wound does penetrate the dermis and if it is so wide that its edges are not easily brought together, then the connective tissue of the dermis can grow in the space between the cut edges of epidermis. We see the healed area as a scar, since the connective tissue is different in appearance from that of the epidermis.

WILL HUMANS EVENTUALLY ALL HAVE THE SAME SKIN COLOUR?

This depends upon how much humans migrate and sexually commingle in the future. Skin colour is determined by the genes you inherit. The pool of genes from which your particular genes are selected is largely determined by the relative amount of in-breeding and out-breeding that has taken place among the population in the geographical area in which you are born. According to Dr Nina Jablonski,[21] people near the equator have developed dark skin in order to block out the sun's harmful UV rays and protect the body's vitamin reserves, especially reserves of folate (vitamin B_9). Without sufficient folate, sperm production is lessened, males are less fertile and children are more likely to be born with defects such as spina bifida. People far from the equator have developed fair skin in order to absorb more sunlight and produce adequate amounts of vitamin D during the long winter months. Without sufficient vitamin D, calcium is lost from bones, causing osteoporosis.

There was less mobility in earlier eras of human history. Now, people move in great numbers all over the world. If in the future there is even more migration between the geographical areas of the world, no particular type of skin colour will be as dominant in any particular region of the world. Technology is affecting this too. Travel is easier and faster, heating and air conditioning are more widely available, vitamin supplements can be taken, and even sunscreen and sunglasses can make living easy in a region to which your skin isn't naturally suited. The world is getting smaller in all sorts of ways.

DOES MASSAGING THE SKIN REALLY DO ANYTHING?

Massage therapy is often dismissed as a spa treat or a mere intimate dalliance with oils by the idle rich and pampered. But recent health research suggests that massage may offer surprising health benefits to old and young alike:

- At the Alfred James Cancer Hospital of Ohio State University, Pauline King and Richard Jost studied the effects of massage on cancer patients experiencing physical discomfort and anxiety. Some subjects received a 15-minute massage of the hands, feet and neck, whilst others just sat with someone. It was found that patients who received the physical contact later reported significantly less pain and stress than did the non-contact patients.[22]

- At the New Jersey Medical School, Diane Zeitlin and colleagues found that massage boosts the immune system in times of anxiety, stress and fatigue. Zeitlin gave nerve-racked medical students an hour-long massage 1 day before a major exam. Afterwards, their blood tests indicated increased numbers and activity of white blood cells and natural killer-cells — immune system defenders that attack viruses and tumours in the body.[23]

At the King's College Hospital of the University of London, Alison Tinsdale conducted a study that found that elderly people who received massages "exhibit less depression and less loneliness, make more social phone calls, visit the doctor less often, and drink less caffeine and alcohol than others [who] did not receive massages". Furthermore, nursing home patients "who get massages frequently display fewer signs of senility". Finally, "agitation [frequently seen in Alzheimer's patients] decreases in Alzheimer's patients who are massaged".[24]

At the University of South Carolina Medical School, Rosemary Smith and colleagues found that women who had recently experienced the death of a child were less depressed after receiving a series of therapeutic massages. The theory behind this effect, they argue, is that massage counters isolation, relaxes depressed people, slows their physiological processes, allows the body to recover naturally, reduces stress hormones and generally feels good.[24]

At the Medical College of Ohio in Toledo, researchers led by Dr Thomas Birk found that AIDS patients who received 45-minute massages once a week had lower levels of stress hormones and more natural killer-cells, which help in fighting the disease. "We don't know if the increase in killer-cells is significant enough to prevent illness," claims Dr Birk, "but we do know that the patients' emotional state improves significantly. And since stress suppresses the immune system, the

better the HIV patient feels, the better equipped
they'll be to ward off infection."[24]

At the University of Maryland School of Medicine in
Baltimore, Dr Gary Hack and colleagues found
evidence that massage works better than aspirin and
other pain relievers in relieving headaches. The team
discovered a set of connective tissues linking neck
muscles to the dura mater, which covers the brain and
spinal cord. According to Dr Hack, "When these
muscles contract, they pull on the membrane, causing
headaches. Since massage relieves muscular tension, it
may be more effective at relieving this type of pain
than aspirin or other analgesics."[24]

Yet it is the findings of the effects of massage on
infants and children that are the most impressive. For
details of these, see Chapter 1.

CAN OUR SKIN "TURN WHITE WITH FRIGHT"?

In this case the cartoons are right. You can turn white with
fright — but only temporarily. In some ways this is the
opposite of turning red from blushing. In blushing, the
vessels of the face expand (dilate) with extra blood, causing
the red colour. Our emotions trigger nerves that in turn
signal the blood vessels to carry more blood. When we laugh
heartily the blood vessels also expand, causing the same red
colour in the face.

However, there are also nerves that contract (constrict)
blood vessels. These are called vasoconstrictor nerves. When
an emotion triggers the vasoconstrictor nerves, they signal
the blood vessels to lower their volume of blood. During the
highly charged emotion of a sudden scare, the vasoconstrictor
nerves become stimulated and cause the blood vessels to
contract to the degree that someone may seem to "turn
white with fright".[25, 26]

WHAT IS FROSTBITE?

Frostbite is a condition where the skin and underlying body tissue freeze. It occurs when body parts, usually the extremities, such as the toes, feet, fingers, ears, nose and cheeks, are exposed to extremely cold conditions. According to the New Zealand Dermatological Society, frostbite rarely occurs in fit and healthy individuals in still air temperatures above −10° Celsius (14° Fahrenheit). But it may occur at higher temperatures if the wind-chill factor is great.

In response to extreme cold, blood flow to the skin and extremities is slowed down as blood vessels constrict (narrow). Constriction occurs so that blood can be redirected to the vital internal organs such as the brain and heart to keep the body warm and alive. Believe it or not, ice crystals actually form in the tissues, the blood vessel walls are damaged and cells start to break down. In response to still further extreme cold, blood vessels briefly dilate (widen) before constricting again. This occurs because the body is trying to preserve as much function in the extremities as possible. However, blood returning to the extremities can leak out through the damaged blood vessels and this can further damage body tissues.

As with burns, frostbite is measured in degrees:

☞ First-degree frostbite (mild) involves the top layer of skin (epidermis) — the skin is white and numb and feels stiff, but the tissue underneath is still warm and soft.

☞ Second-degree frostbite (superficial) involves skin that is being white or blue and feels hard and frozen — blisters usually form with 24 hours and are filled with clear or milky fluid, and tissue underneath is still intact but medical treatment is required to prevent further damage.

☞ Third-degree frostbite (deep) involves white, blotchy, and/or blue skin — the tissue underneath is damaged and feels hard and cold to touch. Blood-filled blisters form thick black scabs over the next days and weeks. Proper medical treatment by personnel trained to deal with severe frostbite is required to help prevent severe or permanent injury. Amputation may be required to prevent severe infection.

☞ Fourth-degree frostbite (full-thickness) damages the skin, muscles, tendons and bones, and results in tissue loss.

WOULD MY CLONE HAVE THE SAME FINGERPRINTS AS I DO?

(Asked by C. Monk of Providence, Rhode Island, USA)

No, it wouldn't. The reason is that fingerprint pattern is not entirely determined by genetics. Rather, fingerprint pattern is determined in part by the pattern of nerve growth into the skin, and this is not exactly the same even in a clone and their donor.

DO WOMEN'S AND MEN'S FINGERPRINTS DIFFER?

Fingerprints differ in women and men. Women's fingerprints tend to have ridges that are smaller and narrower, with more uniform width.

WHY DO I HAVE CREASES ACROSS THE PALM OF MY HAND?

You've asked quite an intriguing question — believe it or not, those lines on your palm are related to fingerprints, and can be used to identify people just as fingerprints can. They are unique to each individual and are called palmar flexion creases. Even identical twins don't have the same patterns of palm creases!

Palmar (or palm) creases form by the third month in utero and, like fingerprints, never change (unless there is permanent scarring of some kind) until the decomposure of the skin at death. Both fingerprints and palm creases are found on the padded side of the hand, a volar area. The bottoms of your feet are also volar areas, and the creases and patterns found on the skin on these parts of the body are types of friction ridges. Since humans originally walked on all fours, we can see that friction ridges once helped humans get a grip on what they were walking on! Even today, imagine the difficulty of holding a slippery glass of water with totally smooth palms and fingers (almost as if you were wearing a slick rubber glove!).

Each human has 3 palmar creases per hand, except in some cases of congenital defects or disease. The crease that loops down around the thumb area is called the longitudinal radial crease. The 2 going partway across the palm are each a type of transverse crease. Some people non-scientifically refer to these 3 creases as the head, heart, and life lines. People with genetic defects often have an abnormal pattern of palm creases; for example, in many Down's syndrome individuals, both transverse creases may be fused together to form 1 long crease across the entire palm.

Forensic scientists sometimes use palm creases when evaluating criminal cases (e.g., a bloody palm print found on a wall). They believe that palm crease patterns are a valuable tool for identifying individuals, just as much as fingerprints are. There is, however, still some debate over the accuracy of this means of identification. The study of skin ridges is called dermatoglyphics, which literally means "skin carvings".

WHERE ARE THE RINGS AND WORMS IN RINGWORM?

Certain infections of the skin, nails, body and hair are commonly known as ringworm. The funny thing is that it is not caused by wearing any kind of ring. Nor does ringworm have anything to do with worms — and it is certainly not caused by worms.

Ringworm is the result of a group of fungi that feed as they penetrate a person's skin. The fungi spread outward in a circle and those in the centre gradually die. This produces the familiar ring pattern on the skin from which ringworm gets half its name. Technically referred to as tinea, ringworm is also called scalp ringworm, jock itch or athlete's foot depending upon where on the body it appears. The ringworm fungi are spread only by direct skin-to-skin contact, by indirect contact with an object or surface that an infected person or animal has touched, or, rarely, by contact with soil.

Before it was known that the disease was the result of fungus, it was thought that the infection was caused by small worms. In fact, *tinea* is Latin for "gnawing worms". The fungi that cause scalp ringworm live in humans and animals. The fungi that cause body ringworm live in humans, animals and soil. The fungi that cause foot and nail ringworm live only in humans. After exposure, it takes up to 14 days for the telltale ringworm symptoms to appear.

WHAT IS THE MOST COMMON FORM OF CANCER?

(Asked by Dan Hemphill of Scarborough, Ontario, Canada)

According to the American Association of Family Physicians (AAFP), skin cancer is the most common form of cancer. More than 1 million people in the US and 150,000 Canadians will be diagnosed with a skin cancer this year. According to the Massachusetts General Hospital's Cancer Center in Boston, 40 to 50 per cent of people in the US who live to age 65 will have skin cancer at least once. The AAFP points out that almost all skin cancers are the result of over-exposure to ultraviolet (UV) rays of the sun and, in recent years, of lights in tanning salons. But skin cancer is also the most curable form of cancer.

There are 2 types of skin cancer. The 1st is non-melanoma. There are many sub-types of non-melanomas, including basal cell carcinomas (BCCs), which are malignancies of the basal cells of the skin's top layer (epidermis). A BCC is a very common skin cancer, particularly on the faces of people with fair skin who are overexposed to the sun. According to the Skin Cancer Foundation in New York City, 800,000 of the 1 million skin cancers in the US each year are BCCs. The squamous cell carcinoma (SCC) is the sister non-melanoma skin cancer of the BCC. The SCC probably accounts for 100,000 of the 1 million skin cancers in the US each year. The BCC, the SCC and other non-melanoma type skin cancers are also the most successfully treated type of cancer.

The 2nd type of skin cancer is melanoma. It is a far more serious, but also a far less common type of skin cancer than non-melanoma. A melanoma is a malignancy of the melanocytes of the skin, the cells involved in the skin's pigmentation process. A melanoma is a particularly aggressive form of skin cancer. It also starts in the epidermis, but quickly grows downwards into the deeper cells of the body, from which it may spread (called metastasis) to another part of the body. Metastasis occurs usually through the blood vessels, lymph channels, or spinal fluid. According to the AAFP, citing figures for 2002 from the American Cancer Society, there were 53,600 diagnosed cases of melanoma in the US and 7400 deaths.

THE "GOLDFINGER EFFECT": IS IT POSSIBLE TO SUFFOCATE IF YOUR ENTIRE BODY SURFACE IS PAINTED?

(Asked by Tom Millford of Kansas City, Missouri, USA)

This question has been around for at least 40 years, ever since it was popularised by the 1964 James Bond movie *Goldfinger*. In the movie, a beautiful but unfaithful mistress is murdered in a rather unusual way by the film's heavy, Auric Goldfinger (*aurum* is the Latin word for "gold", by the way). Her entire body is painted gold and she dies through skin suffocation. James Bond (played by Sean Connery) discovers her body and later describes how such "skin suffocation" has "been known to happen to cabaret singers". He adds that to avoid suffocation, a small patch of skin at the base of the spine should be left unpainted. Thus for decades this idea gripped the public's consciousness. Many people believe that you can die through skin suffocation or, more precisely, through overheating due to your body's inability to "breathe" and cool off.

However, James Bond turns out to be a better secret agent than physiologist. Science says that humans simply cannot die in this way. It is true that, unlike mammals, amphibians breathe through their skin, and could therefore die if their entire body were painted. But humans are not amphibians, reptiles or snakes (with a few notable exceptions). So you will not die of the "*Goldfinger* effect".

But there is some further word on this. In 2002, a team of 6 German dermatologists published an article proving that the human skin gets more of its oxygen from the air than scientists had previously thought. Dr Markus Stucker and his 5 colleagues further report that airborne oxygen penetrates almost 10 times deeper into the skin than formerly believed — ranging from a quarter to almost half a millimetre below the skin's surface. This means that oxygen from the air, rather than from the blood, nourishes the entire outer layer of the skin (epidermis) and some of the underlying layer (dermis) where perspiration glands abound. But before jumping to the fantasy conclusion that the "*Goldfinger* effect" may be possible after all, think again. The Stucker team points out that, even with their findings, prohibiting oxygen intake through the skin still has little effect on internal organs. Death by the "*Goldfinger* effect" is still impossible — shaken not stirred.[27]

WILL WE EVER DEVELOP ARTIFICIAL SKIN?

Each year, about 45,000 Australians undergo skin grafts due to skin cancer. A further 45,000 need grafts due to diabetic ulcers on the skin that are difficult to heal. Another 1000 receive grafts due to burns. According to Dr David Mooney and Dr Antonios Mikos, "[i]n the case of skin, the future is here. The US Food and Drug Administration has already approved a living skin product — and others are now in the regulatory pipeline."[28, 29]

Roughly 16 per cent of our body weight is skin. The body devotes 5 to 8 per cent of its metabolism to the skin's maintenance.

The human body contains between 4.2 and 5.5 square metres (14 to 18 square feet) of skin.

Each square centimetre of human skin consists of about 4 million cells (10 million per square inch), 24 hairs (60 per square inch), 35 oil glands (90 per square inch), 6.1 metres of blood vessels (20 feet per square inch), 246 sweat glands (625 per square inch), 7480 sensory cells (19,000 per square inch), 23,622 melanocytes (pigment cells) (60,000 per square inch), and more than 393 nerve endings (1000 per square inch).

Your armpit is the home of up to about 203,000 bacteria per square centimetre (516,000 per square inch).

CHAPTER 8

THE HAIR & NAILS

According to zoologist Desmond Morris, "There are one hundred and ninety-three living species of monkeys and apes. One hundred and ninety-two of them are covered with hair. The exception is a naked ape self-named *Homo sapiens*."[1]

WHY DOES HAIR ON MY HEAD GROW LONGER THAN HAIR ANYWHERE ELSE ON MY BODY?

(Asked by Rodney Thompson of Cincinnati, Ohio, USA)

Every hair on your body is born, grows and dies following a pattern determined by the growth cycle of the hair follicle, which differs from one part of the body to another. For example, the growth cycle of follicles of head hair allows for longer growth than the growth cycle of follicles of arm hair. The growth cycle of hair follicles consists of 3 stages: the anagen stage is the period of active hair growth; the catagen stage is the period when hair-follicle growth stops and degeneration begins; and the telogen stage is the period of hair-follicle death and shedding.

The length that each hair will actually attain is determined by the amount of time the hair spends in the hair-growing anagen stage. If the hair-follicle growth cycle is programmed for long growth, the hair will be in the anagen stage for a long time before it moves to the catagen stage and stops growing. If the hair-follicle growth cycle is programmed for slow growth, the hair will be short and will spend a shorter time in the anagen stage. Hairs of the scalp are the longest hairs of the body because they spend 4 to 8 years in the anagen stage. The scalp hairs of women and men and the facial hairs of men have plenty of time to grow. That's why it's possible to let the hair of the scalp and beard grow all the way to the ground. By comparison, arm hair, leg hair, eyelashes and hair on most other parts of the body have only a 1- to 6-month period of growth in the anagen stage. They don't get a long time to grow, so they don't grow very long.[2]

IS THERE ANY WAY TO SPEED UP FACIAL HAIR GROWTH?

(Asked by Rodney Thompson of Cincinnati, Ohio, USA)

Teenage boys often ask this question. They want to be able to shave so they can act and look older. Guys, it will happen soon enough. Shaving will become a nuisance for your entire adult life. Over a lifetime, the average male will spend about 3500 hours (146 days) shaving. The rate of hair growth is genetically determined and also affected by other factors, including age, overall health, nutrition, general condition of the skin, supply of natural skin oils, overall vigour of hair follicles, and hormones. The rate of facial hair growth and thickness has to do with levels of the male hormone testosterone. The only way to speed up hair growth is to boost one's testosterone level. However, too much artificial boosting of testosterone can result in testicular atrophy, where a male's testes get smaller and sperm production drops. Most males would most likely regard this as too high a price to pay for the earlier and thicker appearance of facial hair.

CAN HAIR "FEEL"?

(Asked by Morgan Thring of Bellevue Hill, New South Wales)

Human hair itself cannot feel anything, because to be able to feel one must have the capability of processing information. Hair would need a nervous system for that, and it doesn't have one. However, hair does have sensory nerve fibres. These transmit information concerning the motion of the hair to the brain. The brain then analyses this information, compares it with your database of sensory information, and interprets the information accordingly. That's why it hurts when your hair is pulled.

DOES MY HAIR GROW BACK FASTER THE MORE OFTEN I SHAVE?

Urban myth alert! No. Hair is dead by the time it's high enough on the surface of the skin to be cut. This is why cutting a hair doesn't hurt. But plucking a hair? Ouch!

HOW MANY HAIRS DO I LOSE EACH DAY?

Most people think they lose only about 10 to 20 hairs each day. But that's way off. According to Dr Diana Tran and Dr Rodney Sinclair of St Vincent's Hospital in Melbourne, it "is normal to lose 50 to 200 hairs each day".[3] Age, level of stress and overall health are some of the factors that influence how many hairs we lose. Each scalp hair grows for at least 4 years before it is shed. Scalp hair grows at a rate of about 1 centimetre (less than half an inch) per month. Forty million of the 147 million males in the US suffer from some degree of hair loss. Twenty million females suffer from it.

Also, it's an urban myth that hair falls out faster if you spend too long under the shower. If this were true, the cleanest people would be the baldest.

WHO HAS MORE HAIR: BLONDES, BRUNETTES OR REDHEADS?

The average human head has about 100,000 hairs. But natural blondes have more hairs on their heads than any shade of brunette or redhead. A blonde has about 120,000 hairs, a brunette about 100,000 and a redhead about 80,000. No one knows for sure why there is this difference.

DOES EVERYONE HAVE DARK EYELASHES REGARDLESS OF HAIR COLOUR ELSEWHERE?

(Asked by Sonja Conners from Lake Placid, New York, USA)

This is a myth that has probably developed from the fact that most people have darkish hair and that mascara is so widely used by women and girls. However, the eyelashes, eyebrows and all the hair elsewhere on the body are generally the same colour. There is some variation in hair colour based upon the region of the body in which it occurs. For example, the facial hair of men is generally a darker shade of their basic colour, whatever that colour might be. In addition, the colour of axillary (underarm) hair may vary slightly from the colour of pubic hair. And both may vary slightly from that of the scalp. Genetics determines hair colour. If a person has blond eyebrows, then they will have various shades of blond hair elsewhere on their body.

CAN HAIR TURN WHITE OVERNIGHT AS THE RESULT OF SHOCK?

Just like in the cartoons? There is no scientific evidence that hair can turn white overnight due to a traumatic experience. However, legend has it that both Thomas More and Marie Antoinette suffered a hair-colour change to white the night before their executions. Some maintain even today that a condition called alopecia areata can turn hair white overnight. But this condition refers to hair *loss*, not hair colour change. Dr Douglas Nelson of Averon-Bergelle, France, described the case of a 45-year-old French farmer whose hair reportedly went from black to white in 14 days. It stayed that way for about 6 months. Then, over a period of 4 months, it grew back to full black as before. He was in perfect health. There was no illness. There was no shock. There is no explanation.[4-6]

WHY DO WE HAVE EYEBROWS?

(Asked by E. Perrins of Eastwood, New South Wales)

Eyebrows serve the purpose of keeping the sweat of our foreheads out of our eyes. Sweat stings, irritates the eyes and impedes vision. Eyebrows also keep rain from rolling down our forehead and into the eyes. This function is helped by the slightly arched shape of eyebrows, which channels moisture around the eyes rather than into them. Eyebrows may also help reduce glare. Other primates have eyebrows, but none have them against bare skin. Eyebrows figure in our body language. Eyebrows are among the most expressive features of our face. A frown can indicate displeasure. A single raised eyebrow can signal scepticism. Two raised eyebrows can reveal surprise. Furrowed eyebrows can depict thoughtfulness. Sometimes such a facial movement can communicate a message just as well as or even better than words.[7]

WHY DO WE HAVE EYELASHES?

Eyelashes serve the purpose of keeping dirt, dust and other foreign objects out of the eyes. Eyelashes and eyelids work together to help keep eyes lubricated and clean by distributing tears evenly across the eyes. Secretions from special glands (zeis glands), located along the edge of the eyelids are also evenly spread across the eyes by eyelashes, working together with eyelids. Eyelashes also help vision.

When you squint your eyes, the hairs of the upper and lower eyelashes actually mesh together and allow you to see better. Of course, eyelashes play a role in flirtation and courtship too. The batting of an eyelash by a beautiful woman expresses a romantic meaning in a language all its own.[7]

ARE THERE SPECIAL CREATURES THAT LIVE ONLY ON OUR EYELASHES?

The demodicid (*demodex folliculorum*) is a tiny mite less than 0.4 millimetre long. It lives in the pores of the skin and hair follicles, and most often in the roots of eyelashes. The mites live their lives standing on their heads and burrowing down into the skin, where they feed on body secretions and dead skin cells. A single female may lay up to 25 eggs around a single follicle. As the brood grows, they sometimes crowd each other so tightly that under a microscope they look like a bunch of bananas. If too many mites occupy a single follicle, they may force the hair to fall out. The demodicid is harmless and does not transmit disease. It sounds more fearsome than it is.[7]

WHY ARE HUMANS NEARLY BALD?
(Asked by Barbara Sansome of Gosford, New South Wales)

We humans have hardly any hair compared with other mammals. Why? Scientists have presented many theories. We discussed these in *The Odd Body 2* and the list of theories continues to grow. Before there were seven, now there are eight. Maybe by *The Odd Body 4* there will be nine. Who knows? By *The Odd Body 5* there might even be ten. It's enough to pull your hair out.

The sexual selection theory
The great Charles Darwin himself argued that most of the human body has no hair in order to accentuate the pubic

regions that do. The more hairless the body, the more the genitals will stand out visually. Darwin believed this provides a better basis of attraction to the opposite sex and hence a greater chance for reproductive and ultimately evolutionary success. Also, this contrast is a form of sexual stimulation that also ultimately plays a role in natural selection.[8]

The body-cooling theory

This theory argues that hairlessness is somehow associated with temperature regulation in a tropical environment. As human groups evolved, most groups lived within 15 degrees north or south of the equator; invariably, conditions were hot. It is argued that hairlessness may be an adaptation to allow rapid heat loss, made necessary because humans, unlike most mammals, hunt during the day and not at night. Hairlessness prevents overheating during hunting chases. Such hairlessness is necessary because hunting chases were activities for which early humans were otherwise poorly adapted physically.

The brain-cooling theory

This view holds that the evolution of the human brain to its present size is incompatible with the preservation of a luxuriant hairy coat. A bigger brain made it possible for early humans to run in the midday sun, at a time of day when most predators seek shade and water and refrain from the pursuit of game. Having extra brain cells, the brain of early humans was less likely to break down while experiencing heat stress during long-distance running. But a heavy, hairy coat would have kept the body needlessly warm and, indeed, damaged brain cells, resulting in disorientation, stroke or even death. Thus, the bigger brain won out over the thicker coat by natural selection. Although we actually have as many hair follicles as the great apes, the hairs that emerge are too thin and short to form a coat.

The defective gene theory

Another theory is that our loss of body hair can only be ascribed to a mutation involving a defective gene. This mutation supposedly swept through the human population early in our hominid evolution, since all ethnic groups in all regions of the world are affected.

The parasitic defence theory

This notion is based on the assumption that it is easier to rid a nearly hairless body of parasites (body lice, ticks etc) than a hairy one. Parasites bring disease and discomfort. The theory goes that the fewer the parasites, the healthier and happier we are.

The clothing theory

According to this idea, with the advent of clothing, it is reasonable to assume that a thick coat of hair was simply no longer necessary for human survival. Therefore, as hair became no longer necessary (vestigial), it wasted away over time. Thus, heavy body hair is to the outer body what the appendix is to the inner body.

The aquatic ape theory

This highly controversial theory claims that humans evolved from a water environment. As such, humans lost their hair for the same reason that whales, dolphins and manatees lost theirs. A fairly large aquatic mammal needs to keep warm in the water. It is better served by a layer of fat on the inside of its skin than by a layer of hair on the outside. Of course, the question immediately arises how did the fat get there?[9]

The grasping infant theory

Non-human primates such as baboons constantly forage for food on the ground or climb trees to flee predators. Their infants must cling to them or perish. A mother's long body hair makes this easier. But as humans evolved, they became more sedentary, foraged less and moved down from the trees. Their infants needed to cling less and body hair became less essential for an infant's survival.

Needless to say, critics of each of the above theories abound. For example, with respect to "the body cooling theory", it has often been pointed out that the human skin, with its abundance of sweat glands, is an excellent heat exchanger — with or without hair. As the late Columbia University anthropologist Marvin Harris wrote, "We cool ourselves by wetting our skin with moisture exuded by our sweat glands. Humans have as many as five million of these glands — far more than any other mammal."[10] While being hairless and hence cooler is advantageous at the equator, it may not be so at the poles; after all, Inuit (Eskimos) are perhaps the most hairless people on Earth. And as for "the parasites defence theory", non-human primates often use "de-lousing" each other as a pro-social activity that restores harmony to a group. Body parasites may be a personal nuisance, but they cement social solidarity too. As Charles Darwin wrote, "Whether this evil [skin parasite infestation] is of sufficient magnitude to have led to the denudation of [the] body through natural selection may be doubted, since none of the many quadrupeds inhabiting the Tropics have, as far as I know, acquired any specialised means of relief."[8]

Which theory is right? Scientists can't decide. They're pulling out their hair wondering.[1, 11, 13]

WHAT CAUSES PEOPLE TO GO BALD?

(Asked by Matt Halliday of Turramurra, New South Wales)

Common baldness is called androgenetic alopecia. It is patterned, progressive and genetically determined. Baldness occurs when the circulation of hormones (androgens) causes hair follicles to shrink. This means that the follicles are unable to support hair growth. The life of a hair is about 4 to 8 years, so it takes several years for all the hair to fall out, all the follicles to shut down and balding to be completed.

One theory about how baldness occurs has to do with sebum, the oily secretion of the sebaceous glands located in hair follicles. When excessive sebum is produced, it can clog the hair follicles. It is believed that this clogging interferes with healthy follicle functioning and ultimately promotes hair loss. Some degree of baldness will be experienced by 96 per cent of Australian men; this percentage is somewhat less for women. Sixty per cent of men will have "significant hair loss by the age of 50"; 50 per cent of women will have "significant hair loss by the age of 60".

Sadly, hair loss can be devastating. As hair diminishes, so does one's self-esteem. According to Dr Diana Tran and Dr Rodney Sinclair of St Vincent's Hospital in Melbourne, "premature or accelerated hair loss can engender considerable negative thoughts and anxiety associated with feelings of diminished attractiveness". They add that "society places importance on physical attributes, especially the appearance of our hair". So when one loses one's hair, it is almost as soul-destroying as losing a limb. Although common baldness (androgenetic alopecia) "is a normal physiological process" and has no health effects, society discriminates against those who experience it.[14]

Echoing this sentiment, Dr Pamela Wells and colleagues from Goldsmith College in London observe that "fortune does not favour the bald". They point out that research shows that "bald and balding men are generally rated less favourably on dimensions such as physical attractiveness, self-assertiveness, social attractiveness, personal likeability and life success". Their study found that "Increasing degrees of hair loss were associated with loss of self-esteem, depression, introversion, neuroticism and feeling unattractive." They add that the younger a man is when he suffers baldness, the greater the psychological impact.[15]

How strong is this impact? A study by Dr Charles Gosselin of Harvard Medical School found that the personality and behavioural changes brought on by premature baldness in particular can be not only serious, but also permanent.[16] Baldness can be a factor in our political system too. In a study by Dr Lawrence Sigelman and colleagues from Yale University, it was found that "bald men were under-represented in American high elective office: the number of bald and balding members of Congress fell significantly short of what would be expected of a randomly selected group of men of the same age".[17] As for American Presidents over the past half-century, all had considerable hair except Dwight Eisenhower. He was first elected in 1952 and re-elected in 1956. His opponent on both occasions was the bald Adlai Stevenson. Gerald Ford had little hair too, but he was never elected. Although there is no Australian study on this point, Australia has had only two balding Prime Ministers — Billy McMahon and John Howard — in the last half-century.[18]

HOW CAN I PREVENT GOING BALD?

(Asked by Matt Halliday of Turramurra, New South Wales)

See your family physician for this one — not a mere anthropologist like *moi*. The short answer to this is to stay healthy and practise good hair and scalp hygiene. Recent pharmaceuticals have proved to be effective for some men in halting and sometimes reversing the balding process. For example, Finasteride is a prescription drug originally developed to shrink the prostate gland in men with prostate problems. As a baldness treatment, it works by blocking 5-alpha reductase, the enzyme that synthesises dehydrotestosterone (DHT). DHT is believed by some experts to trigger the shutdown of hair follicles. Thus, when DHT synthesis is prevented, balding is prevented. Some prescription and natural products are available that reduce sebum production, thus thwarting the balding process.

Of course, the holy grail of baldness research is finding the gene that causes baldness and creating gene therapy that corrects it. According to one team of researchers, we're about halfway there. Dr Angela Christiano[19] and a team of dermatologists report that they have found a gene that governs baldness. They add that their finding could lead to a gene therapy treatment for people who want more hair. The new gene, appropriately called "hairless", is located on chromosome 8 and "goes a long way towards regulating the human hair cycle". According to Dr Christiano, "The discovery of this new gene gives us endless possibilities that may allow us to effectively treat hair loss and possibly baldness within the next five years. It is now within our reach to design ways to grow hair, remove hair, even dye hair genetically and, best yet, this can all be accomplished topically, reducing possible side-effects."[20–21]

CAN YOU GET CANCER OF THE HAIR?

Not of the hair itself, but of hair follicles. A merkel cell carcinoma (MCC) is also called a neuroendocrine cancer of the skin. It is a rare type of skin cancer in which malignant cells are found on or just beneath the skin and in hair follicles.

Twenty-two per cent of American men sport a moustache. Facial hair on men constantly goes in and out of fashion.

❧ ❧

Hoo Sateow of Chang Mai, Thailand, has the longest head of hair in the world. His locks measure 5.15 metres (16 feet, 11 inches). His brother Yee's head of hair is 4.87 metres (16 feet).

Why does the *Mona Lisa* have no eyebrows? It was the fashion during the Renaissance in Italy for women to shave them off.

WHAT CAN FINGERNAILS AND TOENAILS REVEAL?

(Asked by Chantal Liebert of Oceanside, California, USA)

Fingernails and toenails can reveal all sorts of things about our general state of health. It has even been said that, just as our eyes are a window to our soul, nails are a window to our health. Changes in the look, shape or colour of nails can indicate diseases and disorders long before other symptoms show up. Nails are made of protein, keratin and (here's where many are surprised) sulphur. Fingernails grow at about 0.05 to 1.2 millimetres a week. Toenails grow somewhat more slowly. Nail differences or abnormalities are often the outcome of nutritional deficiencies or disorders. White fingernails and toenails can indicate anaemia or kidney problems. Pitted brown spots or splits in fingernail and toenail tips may indicate psoriasis. White nails with pink near the tops can be a sign of cirrhosis of the liver. White lines across the nail may also indicate liver disease. A half white nail with dark spots on the tip could indicate a kidney disorder. Abnormally thick nails might be due to poor blood circulation in the vascular system. Yellow nails can mean you have problems with your liver, diabetes, respiratory disorders or problems with your lymphatic system. Yellow nails can occur many years before the disorder shows up. Dark nails that are flat or thin can be a sign of a vitamin B_{12} deficiency.

Brittle nails are a sign of iron deficiency and thyroid problems. Nails that are very bendable can be a sign of rheumatoid arthritis. Very deep blue nail beds can indicate pulmonary obstruction or even emphysema. Nails that crack, peel or chip easily may indicate that more protein and minerals are needed in the diet. Brittle and soft nails with a shine and with no moon may indicate an overactive thyroid. Nails that separate from the nail bed could also indicate a thyroid disorder. Ridges in the nail could mean an infection or thyroid disorder. Nails that are like a bumpy road can indicate a thyroid disorder too. Nails that resemble hammered brass may indicate a tendency toward partial or total hair loss. Flat nails can indicate Raynaud's disease. This disease affects the circulatory system and result in hands and feet that are continually cold. Unusually wide nails that are square can mean a hormonal disorder. Red skin at the very bottom of the nail bed can indicate a connective tissue disorder. Changes in the fingernails or toenails may signify disorders somewhere else in the body. See a doctor if *any* of the above concerns you. These signs *may* indicate the problems listed above, but not always.

WHY DO SOME PEOPLE HAVE "SPOON" SHAPED FINGERNAILS?

The medical term for this condition is koilonychia. It is defined as a dystrophy of the fingernails. A dystrophy is any disorder arising from a defect. In koilonychia, the fingernails can take on a more rounded, "clubbed" or "spoon" shaped appearance. The nails can also be thin and have raised edges. Koilonychia is sometimes associated with iron deficiency anaemia. According to Dr L. Bry,[22] koilonychia can also be caused by medical conditions such as rheumatic fever, some sexually transmitted diseases and lichen planus (a non-infectious inflammatory disorder of the skin that can affect the nails). The condition occurs most commonly in individuals who cannot oxygenate their tissues well. Conditions such as congestive heart failure, cystic fibrosis, chronic severe asthma or anything else that decreases the ability of the lungs to oxygenate the blood can cause "spoon" shaped fingernails. Although the exact mechanism by which the fingernails (and fingertips themselves) become "clubbed" is not known, poor oxygenation probably adversely affects the growth of the nail plates, thus causing the deformation.

Nails on longer digits grow faster than nails on shorter digits. Nails grow faster in the summer than in the winter.

In China, wealthy men used to grow their fingernails to astonishing lengths. This was done for reasons of status. Their nails would be so long that they were kept covered by a silk bag. A wealthy man could thus demonstrate that he had servants to do everything for him — and that meant everything.[23]

A hangnail has nothing to do with a hanging fingernail as the word suggests (or anything else hanging). The term that describes the bit of torn cuticle that is often painful comes from the old Anglo-Saxon words *ang* meaning "pain" and *naegl* meaning "nail", hence *angnaegl* meaning "painful nail".

Although it appears that fingernails and toenails grow outward from the end, they don't. Instead, all growth occurs under the skin at the base of the nail in a continuous process called accretion. In accretion, living cells eliminate non-living matter accumulated outside the cell. As the nail grows, it secretes a soft, lifeless substance (keratin). This forces the previous growth, killed and hardened by the keratin, out from under the cuticle and forward as it dries into a solid plate. The entire nail, mostly hardened dead skin cells, moves along in one continuous piece.[24]

CHAPTER 9

THE SKELETON, BONES & TEETH

The late actor, bon vivant and raconteur Sir Peter Ustinov once observed that "parents are the bones on which children cut their teeth". One could ask, do they do the cutting before or after they receive their weekly allowance?

WHAT ARE BONES?

Bone is the hard, calcified tissue of the skeleton. It consists largely of calcium carbonate, calcium phosphate and gelatine. Within even the hardest parts of bones there are many minute cavities containing living matter, connected by minute blood-carrying canals. It is not true that human bones are solid, dry, brittle and white. Bones are extremely porous, pulsating, blood-soaked living tissue. They are soft, relatively lightweight and actually have a sponge-like interior. If bones were solid, movement would be very difficult because of the tremendous weight bones would have. Bones are not white either. They actually range in colour from beige to light brown. The white, sterile skeleton bones seen in displays have been carefully boiled and cleaned.

A human bone is made up of 4 distinct layers: marrow, cancellous bone, compact bone and periosteum. The centre of the bone is a core of marrow where white blood cells are produced. Marrow is a thick, jelly-like substance. The next layer is a fairly large section of cancellous bone, thick and very porous material full of blood vessels and cells; it resembles a sponge. Despite its appearance, cancellous bone is quite strong. The cancellous bone is surrounded by a wall of hard, calcified material called compact bone. This layer is smooth and conforms with our popular image of bone. Finally, there is a thin layer of skin-like tissue covering the bone called the periosteum. It is a dense membrane containing nerves and blood vessels that nourish the bone.

WHY DOES A BABY HAVE MORE BONES THAN AN ADULT?

A baby is born with about 300 bones. But they are very soft, compressible and weak. If they were harder and did not compress well, the journey through the birth canal would be very, very difficult. Many baby bones consist entirely of cartilage. This is more flexible than bone, but also weaker. As the baby grows into a child then a young adult, bones take on calcium, which greatly strengthens them. Certain bones "fuse" together to form the 206 bones we have as an adult.

WHY DON'T ALL BONES HARDEN AT THE SAME TIME?

(Asked by Peter Martin of Reid, ACT)

The process by which bones become hard through the build-up of calcium is called ossification. (This is not to be confused with the process of becoming an Australian citizen!) It's surprising, but in a growing human foetus, the first bone to start ossifying is the clavicle (collarbone). Most people think it must be the legs, feet, backbone or skull, but they're wrong. According to Dr Diane Kelly,[1] the clavicle hardens first at the two ends, which are connected to the shoulder and the sternum. After birth, the clavicle grows larger in the same way as other bones, by adding cartilage to its ends and ossifying them. But clavicles do not form cartilage at the end next to the sternum until the bone has nearly finished growing. This happens at about age 19 to 20. Some other bones actually stop growing later than the clavicle. For example, parts of the pelvis do not fully ossify until about age 22 to 25. No one knows why it is that any bone adheres to a particular ossification timetable.

HOW CAN BONES GET DISEASES?

The skeletal system is composed of bones. These bones are still immature in children. But in adults, bones suffer from the same types of diseases that affect the body's other systems. These include:

- Bacterial infections. A bone infection is called osteomyelitis.

- Non-infectious inflammation, such as rheumatoid arthritis.

- Degeneration associated with an old injury or normal ageing, called osteoarthritis.

- Metabolic disease. An example of this is gout, a disorder of uric acid metabolism in the body. The disorder can cause painful collections of uric acid crystals in the joints.

- Cancer of the bone marrow (where white and red blood cells are made). This is called leukaemia.

- Cancer of the bone itself. This is called osteosarcoma.

- Inherited diseases that affect bones. One example is achondroplasia (or achondroplastic dwarfism).

- Injury or trauma that produces fractures or breaks.

Another disease of the bone is osteoporosis. This refers to a loss of bone mass over time, leading to weak bones that are more subject to fractures and damage. This disease is most commonly seen in older Caucasian women who have passed menopause. Other than fractures, osteoarthritis and osteoporosis, none of these conditions is seen very often. Bones are located all over the body and bone diseases may be difficult to detect and treat. The limited blood supply to bones enables them to resist treatment, especially antibiotic treatment of infections. Bone diseases are always serious. This is because they are often slow to heal, may reoccur and generally result in immediate discomfort and disability due to the fact that the function of the skeleton is to provide support and assist with motion for a large body mass.

WHAT ARE BONE FRACTURES AND WHAT'S THE DIFFERENCE BETWEEN SIMPLE AND COMPOUND?

(Asked by Mark Thompson of La Perouse, New South Wales)

A fracture is merely any break in bone or cartilage. There are 7 types of fractures. There is a widespread belief that simple and compound fractures have to do with the number of places in which the bone is broken. This is not true. A simple fracture means that the skin is not cut or perforated by the broken bone, but the bone may be broken in 1 or more places. A compound fracture is 1 in which the skin and protective tissue around the bone are perforated by the broken bone. Again, it doesn't make a difference whether the bone is broken in 1 or more places. In a compound fracture, the skin is opened and there is contact between the bone and the air. A more recent term for a compound fracture is an open fracture: the skin is perforated and there is an open wound down to the fracture. Consistent with this is the more recent term for a simple fracture — closed fracture — since the skin is intact, or closed, at the site of the fracture.

A stress fracture is a hairline break in the bone. X-rays usually do not show evidence of stress fractures. An incomplete fracture is the term sometimes used when a bone does not snap completely into 2 or more pieces. A complete fracture is the term used when it does. A comminuted fracture occurs when bone is crushed into many fragments. An impacted fracture happens when 1 bone fragment becomes embedded in another bone fragment.

WHY DO WE HAVE KNEECAPS BUT NOT ELBOWCAPS?

(Asked by Nathan James of South Coogee, New South Wales)

We have kneecaps (patellae) to protect the delicate, non-articulated joints beneath them and to help resist damage during walking and movement. Elbows do not have that much wear and tear on them, so we don't really need them. The elbow and the knee have very different functions. The elbow joint is formed by the trochlea of the humerus, the trochlear notch of the ulna and the head of the radius. It is a synovial (hinge-type) joint, and is responsible for bending (flexion) and extending the forearm. On the other hand, the knee is the largest and most vulnerable joint of the body. It actually consists of three separate joints. The first is intermediate between the kneecap and the patellar surface of the thigh bone (femur); the second is to the side (lateral) between the femur, the lateral crescent-shaped bone (meniscus) and the lateral calf bone (tibia); the third is midway (medial) between the medial rounded outer surface (condyle) of the femur, the medial meniscus and the medial condyle of the tibia. The knee is partly a hinge-type joint and partly a gliding-type joint. It is responsible for flexion and extension, slight medial rotation and lateral rotation in flexed positions. The stability of the knee joint comes from the ligaments and muscles crossing the joint. The patella itself is a sesamoid bone that develops in the tendon of the thigh's quadriceps femoris muscle. A sesamoid bone is a bone formed inside a tendon where it passes over a joint. Open sesamoid!

HOW CAN YOU TELL A MALE SKELETON FROM A FEMALE SKELETON?

There are some telltale signs to look for when determining whether a skeleton is male or female. The skeleton's overall bone weight is heavier in a male than in a female. All bones have subtle bone markings; in general, bone markings are more prominent in a male. The male skull is heavier and rougher; the female skull is lighter and smoother. The male forehead is usually more sloping; the female forehead is usually more vertical. The male sinuses tend to be larger. The male cranium (the bones of the skull that contain the brain) is about 10 per cent larger than that of the female. The male mandible (jawbone) is larger, heavier and stronger. The male teeth tend to be larger than those of the female. The male pelvis is narrower, heavier and rougher; the female pelvis is broader, lighter and smoother. The male superior pelvic aperture (the upper opening of the pelvis) is heart-shaped; the female superior pelvic aperture is oval-shaped. The male iliac fossa is deep; the female iliac fossa is shallow. The iliac fossa is the smooth inner surface of the ilium, the largest part of the bony pelvic girdle (formed by the hip bones and sacrum), to which the lower limbs are attached. The male pelvic ilium also extends considerably above the sacrum. The sacrum is the triangular-shaped bone lying between the 5th lumbar vertebra and the coccyx (tailbone). The male sacrum is long, with a very pronounced curvature. The angle under the male pubic symphysis (bone merging) is less than 90 degrees, while in the female it is more than 90 degrees. The male coccyx points to the front (anteriorly); the female coccyx points to the back (inferiorly). As in so much else about us, when it comes to the tailbone, males and females point in opposite directions.

CAN YOU SHRINK A SKELETON THE SAME WAY YOU CAN SHRINK A SKULL?

(Asked by Nathan James of South Coogee, New South Wales)

In *The Odd Body*, we discussed how headhunters shrink a skull — a nice dinner party topic! There is no way to shrink an entire skeleton. The famous shrunken heads of Jivaro, Tsantsa, Shuar and other Amazon peoples actually have the skull bones and other parts removed to permit what is left over to be shrunk. According to Dr Paul Odgren,[2] when it comes to the entire skeleton, you would have to leach out the mineral part of bones with vinegar or with a chemical called EDTA (ethylenediaminetetraacetic acid). That would leave you with partially tough, yet also partially soft and flexible bones — but still at their original size.

WHAT ARE FLAT FEET?

The foot is the body's shock absorber. We place up to 5 times our body weight on the foot every time we take a step, depending upon whether we are walking, running or jumping. If the foot didn't absorb the shock, the force of each step would eventually fracture or dislocate the bones of the foot, leg and lower back.

The arch of the foot is located between the ball of the foot and the heel. In normal feet, the arch is high and can absorb heavy shock. A flat foot doesn't have a high enough arch for the foot to properly perform its shock absorber role. Flat feet can often, but not always, cause pain in the foot, shin, knee, thigh, hip and lower back. The normal arch is composed of bones and joints held tightly together by a combination of muscles, ligaments and tendons. When they become too flexible, they lessen their grip on the arch with the result that the arch lowers. Flat feet can be the result of genes, injury, neuromuscular disease, or a weakening of the muscles, ligaments and tendons due to ageing.

Babies are born with flat feet. This is because their arches are not yet formed. The arch normally forms when they begin to walk.

People can have an abnormally high arch, too, and can suffer the same types of pain and problems as people with a low arch.

The most common and damaging problem from flat feet is pronation. ("Pro-nation" — sounds very patriotic, doesn't it?) Pronation is the turning outward of the foot at the ankle so that one has the tendency to walk on the inner border of the foot.

Abnormal walking due to flat feet can cause bunions, hammer toes, calluses, corns and even neuromas (benign tumours of nervous tissue).

Flat feet can also cause heel spur syndrome and plantar fasciitis. Both involve inflammation of the plantar fascia. These are the bowstring-like flat, fibrous tissue layers stretching underneath the sole and attaching to the heel.

A US Army study found that army recruits with flat feet had far fewer training injuries than recruits with normal or high arches.

～～

Some studies suggest that flat feet can be an advantage in certain sports.

DOES BEING "DOUBLE-JOINTED" MEAN I HAVE TWO JOINTS?

(Asked by John Hyland of Highbury, New Zealand)

Being "double-jointed" does not mean that a person has 2 joints. What being "double-jointed" really means is that a person has a higher amount of flexibility in their joints than does a "normal" person. This is what is demonstrated when, for instance, someone, pulls 4 fingers back so far they nearly touch the back of their hand. "Double-jointedness" usually means that the body's connective tissues (ligaments, tendons, muscles, skin) are less taut or less dense than usual, either of which has the effect of making the joint too "stretchy" (called hyperlaxity). Hyperlaxity can occur in many conditions, including Bamatter's syndrome. This rare syndrome combines a disturbance of connective tissues and premature ageing of the body. Symptoms include insufficient calcium in the bones (osteopenia), short stature, growth retardation, poor muscle development (atrophy), a tendency to easily fracture bones, a look of old age even in children, plus hyperlaxity. When Dr Frederic Bamatter and two colleagues first described the syndrome in 1949, they said that their stunted and prematurely aged patients resembled the animated dwarfs in Walt Disney's *Snow White and the Seven Dwarfs* (1937). For a few years the syndrome was called Walt Disney Dwarfism, but it didn't last.

WHAT CAUSES THE RUBBER MAN OF CIRCUS FAME?

(Asked by John Hyland of Highbury, New Zealand)

Hyperlaxity also describes the so-called Rubber Man of circus freak-show fame. He was able to pull his skin and other connective tissues from his body and amaze paying onlookers in a less sensitive era. The Rubber Man may have suffered from Ehlers–Danlos syndrome (EDS). This is a group of genetic-based disorders that mainly affects the skin and joints, but can affect other organs too. EDS results in hyperlaxity of the connective tissues and overall body weakness. People with EDS are born without the ability to produce certain components that are essential to normal connective tissues. As a result, the skin may become loose and the joints unstable. EDS occurs in less than 1 in every 20,000 people. The US National Ehlers–Danlos Foundation has about 2000 members, with its main office in Los Angeles. Their newsletter is called *Loose Connections*.

WHAT IS THE SOUND I HEAR WHEN I CRACK MY KNUCKLES?

(Asked by John Hyland of Highbury, New Zealand)

The "cracking" noise of the knuckles comes from forcing synovial fluid (joint fluid) to pass under pressure rapidly from one side of the joint to the other. When finger bones are suddenly stretched apart, the space between the joint widens, and an air bubble forms in the joint fluid. But this bubble quickly bursts and that causes the sound. Some call it a "crack", others a "pop".[3]

WILL CRACKING MY KNUCKLES CAUSE ARTHRITIS?

(Asked by John Hyland of Highbury, New Zealand)

A joint is a meeting of 2 or more separate bones. Joint fluid bathes joints and provides cushioning between bones so that they avoid grating against each other. Such wear and tear could cause all sorts of problems, including arthritis. It is true that rheumatoid arthritis is characterised by the loss of joint fluid from the joint. But such a fluid loss results from damage to the linings of the joint itself and not cracking.

☞ Dr Peter Bonafede says that "Nature did not intend us to repeatedly stretch the ligaments of the finger joints. I found 2 medical articles that talked about patients who had injured their hands from knuckle cracking. One over-stretched his ligaments and dislocated his fingers. Another partially tore the ligament in her thumb."[4]

☞ A 1990 study of hand function in 200 adults did not find greater rates of arthritis, but did find a greater tendency towards chronically swollen hands among those who cracked their knuckles than among those that didn't.

☞ In a 1999 article, Dr P. Chan[5] and 2 colleagues write that "A question commonly asked of physicians focuses on the possible deleterious effects of knuckle cracking. Patients are usually concerned that the risk of arthritis is increased by the habit; however, reports addressing the potential long-term consequence are controversial."[3, 6, 7]

WHY DO WE NEED TOES?

You might even ask why couldn't humans get along just as well with webbed feet? The short answer is we need toes to walk efficiently. Although the earliest land animals had various numbers of toes (and/or fingers), those that had 5 fingers and toes won out in the struggle for existence. These pentadactyl (5-toed) creatures established the ancestral pattern for all land-dwelling vertebrates. So the real question is not why do we have 5 toes, but why would we have fewer than 5 toes?

In the evolutionary process the basic body plan does not change unless 2 things happen. First, a genetic trait has to appear, through mutation or through a novel recombination of established traits that allows for the change. Therefore, all vertebrates will have 5 toes unless they possess a gene for fewer toes. Second, that gene must somehow confer an advantage on those who have it. Cursorial animals (runners) actually walk on the tips of their toes. It is advantageous for them because it increases their running leverage and allows them to run faster. In other creatures, toes are part of a grasping foot, and that's where more recent human ancestry comes in. Humans are primates, a group of animals that was initially adapted to living in trees. When you are a tree dweller it is advantageous to have a grasping ability on all your limbs. This ability is a defining characteristic of the primate group. When human ancestors left the trees they became more adapted to running on the ground. Part of this adaptation expressed itself in shorter toes. These were advantageous because running on the ground was more important to the early human lifestyle than was climbing in the trees.

Our evolutionary history has determined our body design. We have toes because our ancestors had toes. Although human toes have become shorter in our evolutionary history, there is no reason to think that they will disappear completely in the future. Body design changes occur only if a new genetic pattern appears that is advantageous. Given modern culture, modern medicine and modern shoes, there is no reason to expect that there would be any advantage to a gene for fewer toes even if it did appear. Nearly every walking animal has toes. In humans, the hallux (the big toe) is the last lever used to push off the ground during walking. Without that toe, the mechanics of walking would be very difficult — and not very efficient.

WHY DOES MY ARM RISE BY ITSELF AFTER PUSHING AGAINST SOMETHING?

(Asked by Francis Salmeri of Gisborne, New Zealand)

This OBQ often becomes the subject of a classroom experiment. A student stands in a doorway and pushes outwards on the doorframe. When the student steps through the doorway, their arm rises up, seemingly by itself! The student appears to lose control of their arms temporarily and will often be amazed by this. Onlooking students are often amazed too. This is called Kohnstamm's phenomenon.

There are a number of systems that help us to regulate muscle tension. To begin with, just to bend your arm you have to coordinate the tightening of your elbow's flexor muscles and the relaxing of your arm's extensor muscles. You have to have good muscle control even in ordinary circumstances in order to perform this task smoothly. When there is a problem doing so, for instance, if the person suffers from certain neurological disorders (e.g., Parkinson's disease), such muscle movements are jerky and uncontrolled. According to Dr John Morenski,[8] your arm will rise by itself after pushing against something due to the load upon the "abductors" of your arm. These are the muscles that raise your arm from your side. As you increase the effort of pushing against the doorframe, you decrease the action of opposing muscles. When you release pressure from the "abduction", you lose control of your arms because the muscles opposing the exertion have not had enough time to readjust. But don't be alarmed. After a few seconds, everything will be back to normal. At least with *this* school science experiment, the classroom never blows up.

IS THERE SOME TRUTH BEHIND THE MERMAID LEGENDS?

Perhaps there is some truth to the mermaid legends of ancient mariners. Mermaid syndrome is a rare but recognised medical condition today. It is called by many other names too: mermaid deformity, symmelia, symposia, sympus, uromelia, monopodia or, more poetically, sirenomelia, sirenomelia sequence, sirenomelia syndrome or sirenomelus. Note the "siren" in these names. The mythical Sirens tempted the ancient Greek mariners to their doom with their enchanting songs.

Mermaid syndrome is a birth defect in which the child is born with a single lower extremity or with legs that are fused together. The symptoms and physical findings associated with this condition vary greatly from case to case due to the wide range of possible physical deformities that may occur. The fusing of the legs can occur from the hips all the way down to the feet. The feet may only be partially fused, giving a web or fishtail look. The skin often tapers down to the ends of the feet, which may allow the skin to "flap" like a fishtail. The skin of the legs may be very dry, extremely thick and discoloured, and can thus have a fish-scale appearance.

Sufferers of mermaid syndrome cannot walk, but they can swim. Some can swim extremely well, as their lower body can perform rather like the body of a fish. Sufferers sometimes enjoy swimming and remaining near water for the comfort and exercise it brings and for the temporary relief it can give their extremely dry skin.

Mermaid syndrome is extremely rare. It occurs in only 1 in every 15 million people to 1 in every 70,000 people, depending upon which expert ventures an estimate. Dr S. Managoli and 3 colleagues[9] claim that 300 cases have been reported in the medical literature.[10-11] The condition occurs at 3 times the rate in females as it does in males. Only 3 people with the syndrome are confirmed to be alive today, according to a TV report on CNN on 2 June 2005. The CNN report covered the successful surgery in Lima, Peru, of 13-month-old Milagros Cerron, whose surgery was televised on Peruvian national television. Her doctors claim that years of surgery will be needed to correct all the major abnormalities in her internal organs. However, Milagros may be able to walk in 2 years' time. It is reasonable to speculate that ancient sailors might have seen such "half woman, half fish" humans in or near the ocean. It may form a truthful basis for an ancient legend.

WHAT IS RICKETS?

Rickets is simply the under-mineralised bone in growing children caused by vitamin D deficiency. According to Dr Christine Rodda,[12] "at the beginning of the 20th century, it has been estimated that 85 per cent of children in northern hemisphere urban industrialised cities had rickets".[13] Since then, public health initiatives have greatly reduced the incidence of rickets. Symptoms of rickets range from seizures to slow and stunted growth, and swelling of wrists and ankles to softening of skull bones (craniotabes). In the past decade or so, rickets has made a comeback among children who don't get enough sunshine. Too many hours spent watching TV, on the computer and playing video games?

The tallest man ever recorded was Robert Wadlow of Alton, Illinois, USA. At his death in 1940, aged 22, he was 2.71 metres (8 feet, 11 inches) tall.

～～ ～～

The shortest man ever recorded was Gul Mohammed of New Delhi, India, who measured 57 centimetres (22.5 inches) tall when examined by doctors at Ram Manohar Hospital in 1990.

The familiar snapping sound made when the thumb and middle finger are snapped together does not originate in the thumb and middle finger as one might think. The sound is actually produced when the middle finger strikes the base of the thumb. The snap against the thumb merely provides enough speed to the finger so that its impact with the hand creates the sound. When you snap your fingers, you perform a fillip.[31]

"Pinkie" is not a childish term carried over from baby talk that is used to designate the little finger of each hand. The word is derived from the old Dutch word *pinkje*, which simply means "little finger".[31]

An "ectrodactylyte" is the name for a person who has lobster-like hands or feet. The most famous ectrodactylyte was Joseph Carey Merrick (aka John Merrick), the so-called "Elephant Man". (We looked at him in *The Odd Body 2*.)[31]

If you are like most people who are right-handed, your left hand does most of the work (56 per cent) when you type.

ARE MIDGETS AND DWARFS THE SAME?

Midgets and dwarfs are not the same thing. Midgets are very small people. They are generally defined as being less than 1.32 metres (4 feet, 4 inches) in height. But they have normal body and limb proportions. Dwarfs are most commonly people who have experienced a pituitary hormone deficiency. They are disproportionate in build and generally physically deformed. Contrary to popular belief, midgets and dwarfs almost always have children of normal size. This is true even when both parents are midgets or dwarfs.

The words "midget" and "dwarf" are now considered inappropriate, offensive and hurtful. The preferred term for both is "little people". Or even better, perhaps no term is needed at all. After all, we are all in the 1 human family. Remember too that within a tiny body, even a misshapen body, can grow a large and wonderful spirit.

WHAT ARE GROWING PAINS?

(Asked by Peter Martin of Reid, ACT)

Growing pains occur in young people when their growth is particularly fast. According to Dr Paul Odgren,[14] such pains usually cause soreness or tenderness around the joints, especially the knees.

To understand the cause of growing pains, try to picture what is going on in a growing joint. You have bones (hard tissue) that are growing longer; attached to the bone are tendons (soft tissue), muscles (more soft tissue) and other growing bones. You also have a kind of protective capsule surrounding the joint (still more soft tissue). All that soft tissue has to stretch out to accommodate the longer bones. It is this incredible stretching that causes the discomfort. In a sense, growing pains are like a mild sprain. In a sprain, the same tissues get a little inflamed and tender. Cold weather can add an extra level of stiffness and make things even more painful. If the growing pains are really severe, a physician may advise the use of anti-inflammatory drugs, braces or bedrest. Usually growing pains get better by themselves once the body tissues adapt to the new bone lengths. But try telling that to an active teenager in pain!

IS THERE A LIMIT TO HOW TALL A HUMAN CAN GROW?

The mechanism controlling human height involves the interplay between genetics and the pituitary gland, plus the quality of nutrition and overall health of the individual. The anterior pituitary gland is located at the base of the brain and produces a compound called somatotropin or human growth hormone (hGH). This hormone stimulates the liver to produce several small peptides called somatomedins, or insulin-like growth factors (IGF). The coordinated activities of hGH and IGF result in bone and muscle growth and increased cartilage formation. The lack of a specific IGF called somatomedin C has been shown to be the probable cause of short stature. This is what happens, for example, in pygmy peoples found throughout the world.

Also important in the overall height of human individuals is the timing of puberty. Gonadal steroids (testosterone and oestrogen) function to speed the ossification of the epiphyseal plates at the ends of the long bones of the arms and legs. This limits long bone length and overall height.

It seems that nutritional deficiencies and general health are closely linked to growth rates and final height too. It is believed that non-genetic variables have as much as a 10 per cent influence on final height. For example, the comparatively tall Masai of east Africa have a diet high in protein, including meat, milk and blood. According to Dr David Mooney, an anthropologist from the University of Michigan in Ann Arbor, these provide the nutritional building blocks for their tall stature. Of course, there have been instances of unusually tall people throughout history. These individuals usually have some form of endocrine disorder. They have an excessive hGH synthesis and release.

Theoretically, it may be possible for humans to attain a height of 4.5 metres (15 feet). We have the example of the male giraffe, which can be 5.5 metres (18 feet) high.

Obviously, the mammalian skeleton and cardiovascular system would have to be greatly modified to support such a height. However, the "evolutionary pressure" to attain such genetic results is lacking. Tall people often complain that there are many difficulties for them that other humans do not experience — finding seats that are comfortable, clothes that fit etc. Not everyone dreams of playing professional basketball! According to Dr Michael Dougherty, extremely tall humans would possibly be more accident-prone as well, since they could easily get in their own way. Nevertheless, some scientists believe that if humans ever live on planets with significantly less gravity than earth, it may be possible to evolve taller bodies.

WHAT IS PHANTOM LIMB SYNDROME?

We briefly discussed this in *The Odd Body*. In 70 per cent of amputees, there's a continuation of sensation in the lost arm or leg. Some amputees even believe their lost limb still exists. It's called phantom limb syndrome (PLS). When an amputation of an arm or a leg occurs, it's normally expected that the person will no longer feel sensation in the remaining stump.

PLS came to the attention of medical doctors in a major way when about 60,000 amputees returned from World War I. Originally it was called "hallucinated limb". PLS was famously described in 1940 by Dr S. Feldman[16] in a classic article in the *American Journal of Psychology*.[17] Among other things, Dr Feldman pointed out that it is often the case in double amputees (who have lost more than 1 limb) that the sensations are experienced in both missing limbs. In 1992, Dr Ronald Melzack claimed that PLS occurs in up to 70 per cent of amputees. The sensation has been described as often painful cramping or shooting. It can vary from occasional and mild to continuous and severe. Besides pain, the range of sensations can include pressure, warmth, cold, itchiness and sweatiness. The sensations usually start soon after the amputation, but sometimes appear weeks, months or even years later.[18]

Phantom limbs seem so lifelike to an amputee that some may attempt to step on a phantom leg and try to walk. When a person with a phantom leg stands up, the phantom leg seems to hang down. It unfolds and stretches out when its owner reclines. It bends properly when its owner sits. When walking, the phantom arm swings in perfect coordination. If a person had a tight ring on a finger or a painful bunion on the foot prior to the amputation of the arm or leg, the pain is still felt after the amputation. PLS can also afflict non-amputees. For example, PLS can be experienced if the shoulder is badly crushed in an accident so that the nerves serving the arm are ripped from the spinal cord. Even though the limb remains attached, it experiences no direct sensations. Surgical removal of the limb does not remove the phantom limb sensation.

In cases involving paraplegics, where the spinal cord itself is severed, the limbs, genitals and other body parts may become phantoms. Phantom legs may seem to move, even to the point of physical exhaustion, and even though the person sees plainly that the legs are immobile. It is known that the area in the spinal cord assigned to relaying sensations from the genitals to the brain is located next to the area relaying sensations from the feet. Thus, genital stimulation of people who have lost a foot sometimes triggers sensations in the phantom leg. In 1 reported case, a man experienced orgasms in his phantom leg as well as in his genitals. In another case, by simply stroking a man who had lost 1 arm, the man reported 2 "virtual hands" which were "hidden in his face and shoulder".

There are paralysed people with PLS who cannot recognise that they are paralysed. For example, if someone is paralysed on their left side, even when they obviously fail to pick up objects with their left arm or cannot use their left arm to tie their shoes, they are emphatic that they are not paralysed. A fascinating footnote to this is that when water is squirted into the left ear, many such PLS sufferers are brought back to reality, lose their PLS sensations and admit their paralysis, but only temporarily. In PLS, it's almost as if the missing body part has a mind of its own.

There are now 3 explanations for PLS. Dr Melzack writes, "the oldest explanation for phantom limbs and their associated pain is that the remaining nerves in the stump, which grow at the cut end into nodules called neuromas, continue to generate impulses".[18] However, in cases where the interior nerve fibres leading from the neuromas to the spinal cord have been surgically cut, the PLS sensations either do not disappear or disappear temporarily, only to return after a few weeks or months. Thus, the neuromas are not the source of the sensations.

The second explanation is that the source of the sensations lies somewhere in the spinal cord itself. Thus, the spinal cord sends strange signals to the brain. But no evidence of such a mechanism has been found.

The final, most recent and probably best explanation is that the brain itself is the source of the sensations; somewhere in the vast network of the brain's millions of neurons a charade is taking place. This view has been put forward by Dr V.S. Ramachandran, who contends that "when the area of the brain assigned to the lost limb no longer receives sensory input from the area, it begins to react to sensory input arriving at adjoining areas in the brain. In other words, the idle area overhears nearby signals that are being processed and acts upon them in error."[19]

What is the lesson in all of this? PLS demonstrates that the brain fights to restore normality after the body is mutilated through injury. We do not fully understand the origins of all human sensations. There is still much to discover about who and what we are.[20]

WHEN PEOPLE GET LOST, WHY DO THEY WALK IN A CIRCLE?

Nearly everyone has heard that people who are lost in a desert, wilderness, arctic plain or other remote and expansive region will unknowingly walk in circles. For centuries, this circular movement was assumed to be caused by the body. That is, since one leg is usually a little bit shorter than the other leg, this causes an imperceptible turn in 1 direction or the other with each step. As the steps add up, you go off course either to the left or to the right and end up going around in a circle.

However, this is not what really happens. Experiments have shown that a spiral movement is actually a universal property of living matter in random and disoriented motion. When walking, blindfolded subjects invariably spiral to the left or to the right, as do swimmers, car drivers, aeroplane pilots and ship captains. (This does not apply to train drivers. No points for figuring out why.) In fact, all animals exhibit this same spiralling behaviour, including microscopic creatures such as the amoeba. The reason for this is unknown, but it is not because of a shorter leg, since it is evident in motion other than walking. (Amoebas don't have any legs at all.) It has been speculated that the spiralling has to do with our response to the rotation of the Earth, the Earth's electromagnetic field or the direction-finding mechanisms of the brain. But science doesn't know for sure why we lean to the left or lean to the right. Maybe we should ask an amoeba.[21, 22]

WHICH FOOT IS SMARTER, THE RIGHT OR THE LEFT?

(Asked by M. Reichburg of Chicago, Illinois, USA)

This is obviously a trick question, deserving a trick answer. Your right foot and your left foot are just as "smart" as each other. But what you don't know is that both your feet are "smarter" than you. Try this simple experiment. Sit straight up in a chair. Lift your right foot a few centimetres off the floor and make small circles with it in a clockwise direction. While doing this with your foot, draw the number "6" in the air with your right hand. Now watch what your right foot is doing while you are trying to draw. Why can't you control your own right foot? Your right foot is "outsmarting" you! Try this with the left foot and hand. The same thing happens.

WHAT IS THE HARDEST PART OF THE BODY?

(Asked by Amy Francis of Launceston South, Tasmania)

Teeth. Specifically, the enamel on a tooth is the hardest tissue of the body. But teeth are not entirely composed of enamel. Teeth are made up of 3 different hard tissues, dentin, cementum and enamel, surrounding a pulp chamber. The pulp chamber is the core of the tooth and contains connective tissue, blood vessels, nerves, lymphatic vessels and just about everything else you would find in connective tissue throughout the body. The job of the pulp chamber is to supply nutrients for the cells that produce dentin (odontoblasts).

Surrounding the pulp chamber is the dentin layer. Dentin is just a little bit harder than bone, due to its high mineral content. It is 70 per cent inorganic; the inorganic component is primarily hydroxyapatite crystals that are made mostly of calcium and phosphate. About 20 per cent of dentin is made up of collagen type 1. This is a protein common to connective tissue. The final 10 per cent is water.

The second hard-tissue component of a tooth is cementum. Cementum covers the root of the tooth; it lies over the dentin and is about 1 millimetre thick. Cementum is very close to bone in composition. It is 65 per cent mineral material (hydroxyapatite), 23 per cent organic (mainly collagen type 1) and 12 per cent water.

The third hard-tissue component of a tooth is enamel. This covers only the crown portion of the tooth and is what you see when you look into your mouth. Enamel lies over the cementum and is typically 1.5 to 2 millimetres thick. It is made up of 96 per cent mineral material (hydroxyapatite) and only 4 per cent inorganic material and water. There is no collagen in enamel.

Teeth are so hard that they are the last part of the body to decompose after we die, and make the best fossils. Anthropologists love teeth because they can tell you much about a person (age, diet, health etc). Teeth talk. You need teeth to be the hardest part of your body, since the act of eating is the most abrasive action the human body performs — and everyone needs to eat.

CAN SOME FOODS AND DRINKS REALLY DISSOLVE TEETH?

(Asked by Amy Francis of Launceston South, Tasmania)

Many foods and drinks are capable of dissolving teeth. According to Dr Robert Houska,[23] carbonated soft drinks dissolve teeth. He says, "If you read the ingredients on the soft drink bottle or can, you will see that it contains either ascorbic acid or acetic acid or possibly both. These acids can dissolve teeth. Also, the carbonation in soft drinks produces carbonic acid which can also dissolve teeth. Sugar is also present in many soft drinks. It is easily digested by the bacteria that naturally occur in the mouth, but this digestive action produces acid, which eats away at the outer layer of teeth. A 2004 study found that the risk of tooth erosion was 59 per cent higher in 12-year-olds and a huge 220 per cent higher in 14-year-olds when they drank soft drinks.[24] One of the authors of the study, Dr W. Peter Rock of Birmingham University, said that the study suggested soft drinks were "by far the biggest factor in causing dental erosion among teenagers". As for foods, any natural or manufactured food that is high in acid can also attack teeth. For example, citrus fruits have a high acid content. That is why it is a very good idea to brush your teeth after eating an orange.[25]

DOES CHOCOLATE CAUSE OR PREVENT TOOTH DECAY?

The latest view by researchers is that chocolate does not cause dental cavities and may even *prevent* tooth decay. In 2000, Japanese researchers from the University of Osaka reported their discovery that antibacterial agents found in cocoa beans (a major constituent of chocolate) offset the high sugar content of chocolate. Thus, they reasoned that chocolate could be said to reduce cavities more than increase them. However, the researchers were quick to point out that the cavity-fighting agent is most plentiful in the husk of the cocoa bean and that this is the part of the bean less likely to be included during the chocolate-making process. Furthermore, the study was undertaken in rats, not in humans. The researchers went so far as to suggest that the extract from the cocoa bean husk could be included in mouthwash and toothpaste.[26]

According to the American Dental Association, the calcium, phosphate, lipids and protein content of milk chocolate seem to modify the acid production in the mouth that leads to cavities. Also, the simple sugars contained in milk chocolate are less harmful than the complex sugars found in other foods and are quickly dissolved from the tooth surface. Dr Angela Dowden, a London dental health expert, adds that dark chocolate plays an even greater role in reducing dental cavities. While research is continuing on this question, perhaps tomorrow's newspaper will carry a story claiming the opposite is true.

IS IT POSSIBLE TO GROW TEETH IN A TEST TUBE?

(Asked by Amy Francis of Launceston South, Tasmania)

Yes. It has been shown that teeth can be made in a test tube. It is called teeth tissue engineering. A study headed by Dr Paul Sharpe, presented at the 2004 meeting of the American Association for the Advancement of Science, showed that harnessing stem cells can re-create the tooth buds that naturally form in the early embryo. Scientists implant these cells into the jaw and let the cells take it from there. Using a person's own stem cells eliminates the problem of tissue rejection. Dr Sharpe remarked that "If you can start the [teeth growing] process off, nature will take its course and an organ — in this case, a tooth — will develop as it would in the embryo."[27] Much of the early work in this field was done on the teeth of smaller mammals (mice and rats) and some larger ones (cats and dogs).

To grow a tooth in vitro you need to get foetal tissues or tissues from developing permanent teeth after baby teeth are lost. The cells responsible for the formation of enamel die during the eruption of the tooth into the oral cavity. In 1 experiment, the Sharpe team implanted a growing tooth from a mouse embryo into the mouth of an adult mouse and "switched on" a gene known to be active in growing molars. They simply waited for the tooth to grow — and it did.

There have been some excellent methods developed over the past 2 decades that allow a dentist to implant an artificial tooth that looks so real not even another dentist can tell just by looking that the tooth is artificial.[25, 28]

❧

By age 50, the average person has lost 12 permanent teeth. In the US, more than 100 million people are missing 11 to 15 teeth. By age 70, half of us will be entirely toothless.[25, 28]

∽ ∽

The ancient Romans hammered iron pegs into the jaw to replace missing teeth. Dental implants today are made from porcelain and titanium.[25, 28]

∽ ∽

Tooth decay is the most common non-contagious disease in the world.[25, 28]

WILL WE EVER DEVELOP ARTIFICIAL BONE?

(Asked by David Crook of South Melbourne, Victoria)

Yes. But it's still a few years away. The process is called bone scaffolding. What are called bone morphogenetic proteins (BMPs) are responsible for regenerating human bone tissue. Experiments with BMPs are currently underway around the world. When artificial bone is developed this is what will happen: rather than setting a broken bone in the traditional way, a revolutionary technique will be used, in which a "bone engineer" will merely "expose" damaged bone to BMPs built around a synthetic polymer "scaffold". The "scaffold" will bridge any 2 bone segments and eventually dissolve, leaving healthy, regenerated bone in its place. Bone "paste", similar in composition to real bone, can already be injected into the gaps left by fractures. This eliminates the need for casts and pins. It also yields better results than bone grafts. The paste hardens in 10 minutes and within 12 hours it's as strong as real bone! Bone "paste" has been used successfully throughout the world in patients with fractures of the hip, knee, shoulder and wrist.[29–31]

WILL WE EVER DEVELOP ARTIFICIAL JOINTS?

Where have you been? There are now artificial joints for virtually anywhere in the body. Many are made of stainless steel and generally they consist of 2 parts. A ball-tipped metal shaft is wedged into 1 bone and a cup-shaped metal socket with a plastic lining is fitted into the adjacent bone. The ball plugs into, and swivels within, the socket to restore free, painless movement. There are artificial joints for the shoulder, the elbow, the wrist, the finger, the hip, the knee, the ankle and the toe.[30, 31]

WILL WE EVER DEVELOP ARTIFICIAL CARTILAGE?

What is called "boosted cartilage" is a fascinating development. At Genzyme Tissue Repair in Cambridge, Massachusetts, a technique has been approved by the US Food and Drug Administration that allows doctors to "boost" cartilage cells. The technique could eliminate the need for total joint replacement with a stainless-steel joint in many cases. Healthy cartilage cells are first removed from the joint of a patient — where possible, they are harvested from the affected joint. The cells are then chemically "boosted" and grown in a laboratory for 2 to 3 weeks. Next the cells are injected under a flap of tissue. They develop into normal cartilage that restores nearly frictionless movement in the joint. In the knee, full regeneration takes between 12 and 18 months. Researchers are using a similar approach to regenerate muscle, ligaments and tendons. According to Dr Robert Langer, such "boosted" cartilage can be grown in the shapes of ears, noses and other body parts.[31, 32]

The wrist consists of 8 bones.

There are 1300 nerve endings in every square inch of the hand.

❧ ❧

The ancient Egyptians never smiled in hieroglyphs and in other depictions that survive. Archaeologists say the reason may be that their teeth were horribly rotted. Their diet was highly corrosive to teeth.

❧ ❧

"Scapulamancy" is the method of fortune telling involving the study of cracked shoulder bones.

❧ ❧

The world's shortest married couple were not "General and Mrs Tom Thumb" — Charles and Lavinia Stratton, of P.T. Barnum circus fame. They are a Brazilian couple, Douglas Maistre Brager de Silva and Claudia Pereira Rocha, who were married on 26 October 1998 and measure 89 and 91 centimetres (35 and 36 inches) respectively.

❧ ❧

Sam Stacey, a young woman from Stainforth, UK, is reputedly the female with the longest pair of legs. Her legs measured 1.26 metres (49.75 inches) from hip to heel in January 2001. This is the height of an average 10-year-old English child.

❧ ❧

The knee is the most easily injured joint in the human body. In US hospitals each year, 1.4 million patients are admitted with knee problems.

CHAPTER 10

THE HEART, BLOOD & LUNGS

John Lyly wrote, "If all the earth were paper white, and all the sea were ink, 'Twere not enough for me to write as my poor heart doth think." Gene Fowler wrote, "Writing is easy: all you do is sit staring at the blank sheet of paper until the drops of blood form on your forehead." Walter Charles wrote, "Fill your mind with ideas the way you fill your lungs with air."

WHAT CAUSES THE HUMAN HEART TO START BEATING?

(Asked by Jane Steward of Singleton, New South Wales)

During embryonic development, the heart is 1 of the few organs that must function almost as soon as it is formed. The fact that the heart is one of the earliest organs to develop suggests its importance. As a functional organ, the heart begins to beat very early, even before structures such as valves and the muscular walls that divide the heart's chambers (septa) have formed. The human heart begins to beat and pump blood through the embryo around day 22 after impregnation (gestation). There is an electrical stimulus that triggers the muscular part of the heart (myocardium) to contract. The stimulus is myogenic, which means that the contractions arise spontaneously within the myocardium itself and propagate from cell to cell. Input from the central nervous system can modify the heart rate (the frequency of heartbeats), but it does not start the beats. The ability of cardiac myocytes (the cells that comprise the myocardium) to beat is an intrinsic property of these cells. Myocytes removed from the early heart and grown in a laboratory will beat sporadically! And if they become connected to each other they will then begin to beat rhythmically and in unison! Wow!

The initial contractions are peristaltic. That is, they proceed in a wave-like fashion along the length of the heart. Once the heart has matured and the conductive system has developed, the contractions proceed in an orderly and timed sequence through the different chambers. What causes the heart to begin contracting is a cell biology matter, since beating is an intrinsic property of the cardiac myocytes. Certain things are required in order for the heart to beat — for instance, cardiac myocytes must have contractile proteins (such as actin and myosin) that are properly assembled into a scaffold (sarcomere) that allows contraction to take place. These cells must have specialised structures (gap junctions) that allow them to communicate so they can beat together in a synchronised fashion. While it is clear that certain elements must be in place for the heart to beat, it is not yet clear what stimuli are responsible for initiating the contraction that is necessary for the beat. And the beat goes on![1–3]

IS IT TRUE THAT THE YOUNGER YOU ARE THE FASTER YOUR HEART BEATS?

(Asked by Nicole Trudeau of Port Elgin, Ontario, Canada)

The heart rate decreases in older humans. This is probably related to changes in the brain regions that control the heart rather than to changes in the heart itself. The brain controls the heart rate via nerves that enervate the heart and achieve the required "pacemaker" function. One of these nerves is the vagus nerve. It plays a vital role in the body by also enervating the larynx in the throat and the gastrointestinal system. The nerve cells for the vagus nerve are located deep within the medulla of the brain, in a cluster of cells called the motor nucleus of the vagus. These nerve cells in turn are affected by nearby neurons in the nucleus of the brain's tractus solitarius and in the hypothalamus. Both of these brain regions show damage and altered function in older humans, for reasons that are not completely understood. Other influences upon heart rate include hormones, like thyroid hormone. Thyroid hormone also decreases in quantity as we get older. And this too is probably due to changes in the hypothalamus. Beyond these, there may be additional factors that explain why the heart rate slows when we're moving closer to our use-by date.[4]

WHAT KEEPS MY HEART BEATING?

(Asked by Gary Robbins of South Bank, Queensland)

That attractive person walking by, of course! Seriously, in the simplest of terms, what keeps your heart beating is electricity. The heart is an incredibly strong muscle and the most reliable pump nature ever constructed. The size of a large fist, the heart pumps blood continuously through the circulatory system. Each day the average heart beats 100,000 times and pumps about 7500 litres of blood through 120,000 kilometres of blood vessels. In an 80-year lifetime, an average human heart beats almost 3 billion times. It never stops beating until it fails, just like any mechanical pump. If it is not restarted very quickly the result is usually death.

The beating of the heart is controlled by the generation of repeated electrical signals in two specialised regions of the heart. One is called the sionatrial node and the other is the atrial ventricular node. These electrical signals are sent through the heart in a synchronised manner that makes the heart beat (contract and expand). This is what you hear as the "lub-dup" when you listen to your heartbeat. The synchronised contraction and expansion of all the chambers of your heart is critical for life-sustaining blood flow to be maintained to your brain and body. According to Dr Terry Hebert,[5] while electricity controls the continued running of your heart, more technically, the source of the heart's energy is adenosine triphosphate (ATP). ATP is a high-energy molecule generated by the metabolism of the food you eat. Approximately 70 per cent of the energy of the heart at rest is generated by the metabolism of fatty acids, along with some carbohydrates. ATP? Sounds like a fuel additive for your car. For your heart, that's pretty close to what ATP is.[6]

DOES LISTENING TO MUSIC SOOTHE THE HEART?

Yes and no. It depends on the type of music. Research shows that listening to so-called "relaxing" music tends to decrease heart rate. Listening to so-called "non-relaxing" music tends to increase heart rate. This was the finding of Dr Peter Scheufele[7] reported in the *Journal of Behavioral Medicine*.[8]

However, there is great individual variation in the effect of music on people. According to Dr Eric Tardif,[9] what is relaxing music to some may be non-relaxing to others. Classical music such as Mozart may not necessarily relax all those who hear it. Heavy metal, techno, industrial and other recent forms of rock music may not necessarily excite all those who hear them. As to why "non-relaxing" music is more often preferred by younger people, perhaps they are more likely to enjoy the stimulation it gives to their heart.[10]

CAN YOU STILL LIVE IF YOUR HEART STOPS BEATING?

The strange answer to this is "Yes." Doctors in Russia recently reported the case of a man whose heart stopped beating, yet he continued living. The man, Nikolai Mikhalnichuk, had a heart attack due to the emotional shock he experienced when his wife said she wanted a divorce. At his local hospital in Saratov, it was discovered that, although his heart had stopped beating, the heart's vessels strangely kept on pumping blood. He was eventually discharged and now leads a normal life, including engaging in both light exercise and sex. In the 1950s, Swiss doctors testing a new method for taking a cardiograph found that some people have sleeping parts of heart tissue. This was found in 7 people who had suffered strong emotional stress. Mikhalnichuk's case is somewhat different. His heart is entirely asleep. According to Professor Vitali Levitskii of the Institute of Cardiology in Moscow, only 2 similar cases are known. One is from Brazil, the other is from Japan.[11]

WHAT IS CHOLESTEROL?

Cholesterol is an indispensable component of all cells and is vital to cell growth and survival. It is used to produce cell membranes and some hormones and performs other vital functions such as absorbing fats from food. About $1/7$ of the brain is made of cholesterol. All steroid hormones, such as the stress hormone cortisol, the male hormone testosterone and the female hormone oestradiol, are derived from cholesterol.

Cholesterol is an odourless, tasteless, white, fatty, waxy substance found in the blood within all body tissues. Without cholesterol, we would not remain alive. The human body manufactures its own cholesterol in order to survive and adapt to the environment. The body also gets cholesterol from the eating of animal products such as meat, poultry, fish, eggs, butter, cheese and whole milk. Some foods that don't contain animal products may contain trans-fats. These cause the body to make more cholesterol. Foods with saturated fats cause the body to make still more cholesterol. However, excessive cholesterol levels are believed to be associated with coronary heart disease, leading to stroke and heart attack. Hypercholesterolemia is the term for high levels of blood cholesterol. Natural controls on cholesterol levels are usually maintained by receptors in the liver. But when these natural controls are overwhelmed by too much cholesterol, it builds up along arterial walls. This can lead to dangerous arterial blockages.

WHAT IS THE DIFFERENCE BETWEEN GOOD CHOLESTEROL AND BAD CHOLESTEROL?

Cholesterol does not dissolve in the blood. It has to be transported from cells by special carriers called lipoproteins. There are two kinds of lipoproteins. Low-density lipoprotein (LDL) is the cholesterol that can block arteries. LDL is thus known as bad cholesterol. High-density lipoprotein (HDL) is the cholesterol made by the body that carries cholesterol away from arteries. HDL is thus known as good cholesterol. Studies show that high levels of LDL increase heart disease risk and high levels of HDL *decrease* heart disease risk.

WHAT IS THE DIFFERENCE BETWEEN DIASTOLIC AND SYSTOLIC BLOOD PRESSURE?

(Asked by Milos Brantke of Liverpool, New South Wales)

Systolic blood pressure is the maximal pressure within the cardiovascular system as the heart pumps blood into the arteries. Diastolic blood pressure is the minimum in pressure, as happens while the heart rests and fills with blood, prior to the next contraction. There are many mechanisms the body uses to control blood pressure. Actually, the heart itself plays only a minor role in regulating the blood pressure. Most of this regulating work is done by the kidneys and medium-sized arteries called arterioles.

The heart responds to an increased need for blood flow in 2 ways. First, it increases the number of times it beats (heart rate). Second, it increases the strength of each contraction as well as the amount of blood pumped out with each contraction (stroke volume). Nevertheless, it should be kept in mind that these alterations only come about in response to chemical and physiological changes within the body. It is not unusual for the systolic pressure to rise 10 to 20 millimetres of mercury during strenuous exercise. This is the pressure in the system while the heart contracts, pumps faster and pumps a greater volume of blood during exercise. However, diastolic pressure should not vary more than 5 to 10 millimetres of mercury as this represents pressure in the system when the heart is not contracting. An extreme rise in diastolic pressure during exercise could indicate a serious malfunction within the cardiovascular system. "Normal" blood pressure has been set at 120/80 or lower for an adult. The first number (120) is the systolic pressure (as the heart beats). The second number (80) is the diastolic pressure (as the heart relaxes). The full rendering of this number, expressed verbally as "120 over 80", is "120/80 mmHg (millimetres or mercury)" or as "120 mmHg systolic and 80 mmHg diastolic".

CAN YOU STOP YOUR PULSE?

This is an old magic trick. All you have to do is block the artery that goes to your arm and the pulse in your wrist will appear to stop. This trick is usually done by hiding a small ball or wad of cloth in the armpit. But don't try it. Blocking an artery for an extended period of time is not recommended, as it could mean that your tissues do not get enough oxygen.

WHAT IS A HEART ATTACK?

The heart is 1 big muscle that pumps blood to the rest of the organs in the body. The heart muscle itself needs some of this blood to give it the energy to keep pumping, so there are arteries and veins inside the heart muscle that transport blood. But sometimes the arteries that bring blood to the heart muscle can narrow as sticky debris — atherosclerotic plaque — begins to coat the inside of the vessel. Plaque consists of cholesterol, calcium, clotting proteins and other substances. The process of developing atherosclerotic plaque is now known to begin as early as childhood. However, even late in adulthood, lifestyle changes (e.g., better diet, more exercise etc) can reduce the onset or severity of coronary artery disease.

Narrowing of arteries usually isn't a big problem, but sometimes a little blood clot (intracoronary thrombus) can form, or a piece of the plaque can break off. These can travel through those narrow arteries for a little while, but sometimes the arteries get clogged. When this happens, less blood can travel through those vessels to the heart muscle. If the blood flow is significantly reduced, the heart muscle will start to suffocate from the lack of oxygen to this area. Blood carries the oxygen to the muscle. If the muscle suffocates and dies, the heart won't be able to pump very well, if at all. Thus, the rest of the body will not get as much blood as it usually needs.

The heart and the rest of the body try to help out when this starts to happen. According to Dr G. Monreal,[12] the brain sends signals to the heart to tell it to beat faster in order to pump more blood. Cells in the heart release chemicals to try and help the heart beat more strongly. But when there is less oxygen in the muscle, the mitochondria (cell food) in the cells have trouble generating energy. This causes the production of lactic acid. Lactic acid build-up in muscles causes that burning and tingling sensation one can experience after too much exercise. When someone is having a heart attack, they sometimes feel a pain in their chest area that seems to spread to the neck, arms and shoulders. This pain is also thought to be from the lactic acid build-up in the heart muscle. The pain feels like it's "radiating" because the lactic acid can affect other nerve endings in the area. The heart isn't beating very well any more, so it can't pump enough blood to the rest of the body. Blood is still returning to the heart from the veins, so sometimes it backs up in the lungs, waiting for the heart to pump it forward. This can congest the lung tissue and result in rapid and shallow breathing. If the heart isn't pumping well, it can't supply enough blood to the brain and other regions of the body. If the brain doesn't get enough blood, a person can black out and die.

IS THERE A LIMIT TO THE NUMBER OF HEARTBEATS WE CAN HAVE IN A LIFETIME?

(Asked by D. Fortier of Quebec City, Quebec, Canada)

There is a fascinating law in biology: in general, all mammals that manage to reach old age accumulate about 800 million heartbeats, regardless of their size, and then die. For example, a shrew, with a heart rate of almost 1000 beats a minute, dies of old age in about 1.5 years. An elephant, with a heart rate of 30 beats a minute, dies of old age in about 50 years. Both animals expire when the 800-million heartbeat limit is reached. So what about humans? According to the law, humans should only live to about age 25. But our heartbeat limit is 3 or 4 times this much. Humans break the biological law and are the only mammals that do. This fact stumped even the late Stephen J. Gould, the great Harvard paleontologist, who couldn't explain why humans are unique in this regard. He wrote: "But *homo sapiens* is a remarkably deviant mammal in more ways than braininess alone. We live about three times as long as mammals of our body size 'should', but we breathe about three times as often as an average mammal of our body size."[13]

Heart rate in humans varies depending on various factors such as age, physical state and stimuli. A young human has a smaller heart and it needs to beat faster to pump the proper amount of blood through the growing body. The heart rate of an infant is 120 beats per minute, for a child 90 per minute, for someone over 18 about 70 per minute. A physically fit person has a lower heart rate than an inactive person. Stimuli resulting in anxiety, stress, fear or excitement will result in a more rapid heartbeat. Nerves connected to the heart regulate the speed with which the cardiac muscle contracts. In an average lifetime, the human heart continuously beats more than 2.5 billion times — way beyond the 800-million limit.[14]

WHAT IS HEARTBURN?

Heartburn has nothing to do with the heart! Heartburn (pyrosis) is a burning sensation felt behind the breastbone, sometimes even radiating into the neck or back. It is caused by the presence of gastric secretions (called reflux) into the lower portion of the oesophagus near the stomach. It's a backing up of the acids from the stomach. This unpleasant chemical retreat usually enters and leaves its burning residue in our oesophagus. Sometimes the acids travel all the way up to our mouth, causing a hot, bitter or sour liquid taste. This condition is called waterbrash. The reflux, primarily hydrochloric acid, causes pain as it irritates the lining of the oesophagus. Although failure of the upper stomach muscles to contain gastroesophageal reflux is generally considered a minor digestive disorder, heartburn can indicate a more serious disorder of the digestive tract. The name "heartburn" probably came from the fact that the pain often seems to radiate from around the heart. But since most people don't know where their heart is actually located, it's easy to see how confusion could emerge.

The most common reason why people get heartburn is due to overextending their stomach by eating too much food. When the stomach is forced to expand to near its limit, the digestive processes underway are shifted upwards. Juices and gases are more likely to be brought back up because they were higher to begin with and have less distance to travel. When heartburn happens, the back pressure of gastric secretions opens a muscular valve called the cardiac sphincter, located between the bottom of the oesophagus and the top of the stomach. Because gravity is often involved, bending over can sometimes put more pressure on an already sensitive area of the cardiac sphincter, thus increasing the heartburn pain.

A surgical procedure called a laparoscopic fundoplication has been available for more than a decade to help treat serious chronic heartburn. Operating with a laparoscope through small incisions in the abdomen, surgeons wrap the top of the stomach (fundus) around the lower end of the oesophagus. The wrap is intended to strengthen the cardiac sphincter muscle at the top of the stomach so that it opens only when it's supposed to. (A laparoscope is an instrument through which structures within the abdomen and pelvis can be seen. Laparoscope comes from two Greek words, *lapara*, which means "flank", and *skopein*, meaning "to see".)

WHAT IS A HEART BYPASS OPERATION?

Speak to your family physician about this if you have any questions about your health. When a person has coronary artery disease (CAD), a common surgical treatment is a coronary artery bypass graft (CABG), or "heart bypass" for short. CAD is characterised by the hardening, narrowing and blocking of the coronary arteries (atherosclerosis), which supply oxygen-rich blood to the heart. If left untreated, CAD may lead to a heart attack. During a CABG, a surgeon harvests a segment of a healthy blood vessel (either an artery or a vein) from another part of the body and employs it to create a detour, or bypass, around the blocked portion of the coronary artery. A CAD patient may need 1, 2, 3 or 4 bypasses depending upon how many coronary arteries are blocked.

There are medications that can reduce blockage to some degree. Only about 10 per cent of CAD patients need a CABG. There are also medications that decrease the amount of work the heart has to do by lowering the heart rate and blood pressure. Angioplasty may also be an alternative. This is a surgical repair of a blood vessel. Balloon angioplasty is a procedure where a balloon-tipped catheter is introduced into a diseased blood vessel. As the balloon is inflated the vessel opens, which allows improved blood flow. Laser angioplasty involves the use of a laser beam to open the vessel. There are several factors that determine which procedure is the best for a patient. These include the number of vessels that are blocked; the shape, length and location of the blockages within each vessel; the age and health of the patient; and the experience and ability of the doctor performing the procedure.

After angioplasty, the blockage can sometimes reoccur. This tends to happen gradually, most frequently in the first 3 months after the procedure. It occurs occasionally during the second 3-month period and rarely after 1 year. If a vessel doesn't close off after 1 year, it is considered rare that it will close over the next 5 to 10 years. About half of all bypass grafts close off in 10 to 12 years. Thus, if a patient had a quadruple bypass, then 2 vessels would be expected to be closed in 12 years. How well a bypass patient takes care of themselves has a lot to do with how long the grafts will remain open.

COULD THE HUMAN HEART BE ON THE RIGHT SIDE JUST LIKE THE HEART OF MR SPOCK FROM "STAR TREK"?

(Asked by C. Hampton of Scarsdale, New York, USA)

Body asymmetries are such that if the heart switched sides and if the slightly smaller left lung and the slightly larger right lung stayed as they are, there would be problems. The heart would apply too little congestive pressure on the left lung and too much congestive pressure on the right lung. This would result in an overinflation of the left lung and an underinflation of right lung into a larger-than-needed chest cavity. This, along with other congenital defects, can lead to heart failure. Overinflating or underinflating of the 2 lungs can happen in a small number of people. It is a condition called dextrocardia and often requires surgery to correct it. In the James Bond novel and film *Dr No*, the evil Dr Julius No suffered from dextrocardia.[15]

COULD WE LIVE NORMALLY IF OUR ORGANS WERE ON THE OPPOSITE SIDES OF OUR BODY?

(Asked by C. Hampton of Scarsdale, New York, USA)

Recent interest in left–right organ asymmetry has arisen from the discovery of a handful of genes that cause situs inversus (complete reversal of all the internal organs) in humans. It appears that very early in our body's formation, a left–right axis is generated based on the orientations of the head–tail (anterior–posterior) and back–front (dorsal–ventral) axes. In embryonic development, this left–right axis is the expression of different genes. The genes instruct each side of the embryo to develop one kind of tissue to form 1 kind of organ, and another type of tissue to form another kind of organ. Sometimes both sides of the body are instructed to form the same tissue, and thus the same organ is formed on both sides of the body. For example, when the spleen is formed, the cells on the left side of the body have a "left" signal that promotes spleen growth, while the same cells on the right side don't see the signal and so do not form a spleen.[15]

DOES THE HEART HAVE ITS OWN BLOOD SUPPLY?

(Asked by Gary Robbins of South Bank, Queensland)

The muscles of the heart need blood just like any other muscles in the body. Oxygen and nutrients have to be carried along as blood. The walls of blood vessels have to be fairly thick to handle the great distances blood has to travel in the human body. Thin walls are fine if you're the size of a flea, but to pump blood around a larger body, the heart needs much more muscle mass, so much thicker walls. This can only be accomplished by giving the walls of the heart their own capillary "beds" to supply them with oxygen and nutrients through coronary vessels.

The coronary arteries arise from the base of the aorta as it exits the heart and fan out under the visceral pericardium (the outer "skin" of the heart), sending smaller arterioles into the muscle to supply oxygenated blood to the capillaries. The capillaries drain into the coronary veins that join to form the coronary sinus, which carries the used blood directly back to the right atrium, separating the coronary circulation from the systemic circulation and the pulmonary circulation.

Early in the development of the heart a cluster of cells lying behind the heart migrate across the surface of the heart to form a thin membrane called the epicardium. The epicardium then divides into 2 layers. The first is an outer "skin", the visceral pericardium. The second is an inner network of blood vessels called the epicardial vessels. These are connected to the rest of the circulation system at the closest points. They also happen to be the major vessels from the heart. These are the ventral aorta and the common cardinal veins. For many reasons, the epicardial vessels empty almost exclusively into the left common cardinal vein. As the embryo continues to develop, the body uses the right common cardinal vein more than the left, until finally the right common cardinal becomes responsible for the systemic circulation and the left common cardinal becomes responsible for only the coronary circulation. At this point, the right common cardinal vein is called the superior vena cava. The left common cardinal vein is called the coronary sinus. They have access to the right atrium of the heart. Thus, the coronary circulation is separate from both the systemic circulation and the pulmonary circulation.

WHY DOESN'T MY BLOOD FREEZE ON A REALLY COLD DAY?

Blood is composed mostly of water. The freezing point of blood is very close to that of water (0° Celsius, 32° Fahrenheit). Proteins, salts and other constituents of blood alter this temperature slightly. But the body keeps a steady temperature even on days when the temperature is below freezing. Some fish have a glycoprotein that serves as an antifreeze that allows them to swim in frigid waters. Humans don't need this protein. If we lived our lives swimming in the polar oceans, perhaps we would.[16]

HOW MUCH BLOOD IS IN MY BODY?

The average adult has between 4.7 and 5.7 litres (5 and 6 quarts) of blood in their body at any given time.[17]

WHAT IS PLASMA?

Approximately 55 per cent of blood is plasma. Plasma is a yellowish-coloured liquid that is 90 per cent water. It also contains salts and minerals such as calcium, potassium, magnesium and sodium. It carries cells and platelets that stop bleeding and prevent infection. Without plasma, blood cells could not be transported throughout the body.[17]

WHAT ARE PLATELETS?

Platelets are colourless and oddly shaped bodies present in blood. They have a sticky surface, allowing them to join with other substances to form blood clots when bleeding must be stopped and disease must be fought.[17]

WHAT'S THE DIFFERENCE BETWEEN RED AND WHITE BLOOD CELLS?

Blood is alive, since it contains living cells. Red blood cells specialise in transporting oxygen throughout the body. They have high concentrations of haemoglobin. Haemoglobin is a conjugated protein, containing haem groups and globin. Haem groups consist of 4 chains of proteins. Globin consists of 4 polypeptide chains, each consisting of hundreds of amino acids. White blood cells circulate in the blood, lymphatic system, lymph glands and spleen. They are part of the immune system and are responsible for fighting harmful invaders of the body. They do this directly, through the action of T-cells and macrophages, and indirectly, through the action of B-cells that produce antibodies. Macrophages (from the Greek words *macros*, "large", and *phagein*, "to eat") are cells within the tissues that originate from specific white blood cells called monocytes. T-cells and B-cells belong to a group of white blood cells called lymphocytes and play a role in the body's immunity.

Both red blood cells and white blood cells are important in nourishing and cleansing the body. But they too need nourishment, since they are alive. Both red and white blood cells have a definite life cycle, just as all living organisms do.

When the human body loses a small amount of blood through a minor wound, the blood platelets cause the blood to clot so that the bleeding stops. The body itself can then replace the lost blood because new blood is always being manufactured in the interior of the bones. However, when the body loses a large amount of blood through a major wound, a different situation develops. The clotting activity of the blood platelets and the manufacturing of new blood may not be enough to restore a sufficient blood supply. A blood transfusion may then be necessary. Blood may be drawn either from another person or persons with a compatible blood type, or from a previous blood donation of one's own blood that has been stored for an emergency or for future surgery (called autologous blood donation).

IF OUR BLOOD IS RED, WHY ARE OUR VEINS BLUE?

(Asked by E. Perrins of Eastwood, New South Wales)

Blood has a "blue blood" stage. The blood we see in our veins near the skin's surface has been stripped of its oxygen load and looks blue. It is close enough to the surface of the skin for its colour to be seen. Inside the red blood cells that make up about 40 per cent of blood volume is the oxygen-carrying pigment haemoglobin. As the red cells pass through the lungs, haemoglobin picks up oxygen and binds with it. It turns bright red as a result of this process. "Oxyhaemoglobin" is the term used for the product of this oxygen–haemoglobin combination. Oxyhaemoglobin is pumped out of the heart under pressure through large muscular blood vessels called arteries. Red blood cells with their oxyhaemoglobin eventually pass through tiny blood vessels called capillaries, where they give up their oxygen to other cells for use in the body's metabolism. (Metabolism is the sum of all the physical and chemical processes by which any organised living substance is produced, maintained or transformed into energy for any animal, including humans. Metabolism is from the Greek word *metabole* which means "change".) As haemoglobin loses its oxygen, it turns a dark purplish-blue. "Deoxyhaemoglobin" is the term used for the product of this former oxygen–haemoglobin combination. Deoxyhaemoglobin collects in larger and larger veins on its way back to the heart. While the biggest veins are deep in the tissues (they tend to be paired with the largest arteries), some fairly large veins are just under the skin, where their blueness can easily be seen by the naked eye. However, if a cut occurs and blood in the veins is exposed to oxygen in the air, it immediately turns red. That's why blood is red when we see it outside the body. But you could say we're all "blue bloods" to some extent.[17,18]

WHAT IS ANAEMIA?

(Asked by Dan Hemphill of Scarborough, Ontario, Canada)

Briefly, anaemia is a condition in which there are too few red blood cells in the bloodstream. Either red blood cells are lost or they are destroyed faster than they can be replaced. This results in an insufficient oxygen supply to tissues and organs, which means that the body has less energy than it needs to function properly. Anaemia may become worse if it is not treated, and it can lead to potentially serious, even life-threatening complications. When the number of red blood cells decreases, the heart must work harder, pumping more blood to send more oxygen throughout the body. If the heart works too hard, it can develop a rapid heartbeat (tachycardia), and/or another serious condition known as left ventricular hypertrophy (LVH), an enlargement of the heart muscle that in turn can lead to heart failure. According to the Mayo Foundation for Medical Education in Rochester, New York, LVH is characterised by weakness, pale skin, tachycardia, shortness of breath, chest pains, dizziness, cognitive problems, numbness or coldness in your extremities and headache. But these symptoms can be caused by other problems too.

There are approximately 100 types of anaemia. The causes are many, including serious disease, vitamin or iron deficiencies, blood loss, genetic problems or a side effect of medication. The US Centers for Disease Control and Prevention estimate that approximately 3.4 million people in the US have anaemia. Approximately 100,000 to 200,000 Australians have anaemia. See your physician if you have any concerns in this regard.

WHERE IS MY SPLEEN AND WHAT DOES IT DO?

(Asked by Yvonne Chambers of Hackney, South Australia)

The spleen is an organ of major importance to the body's immune system. The spleen produces white blood cells that fight disease and infection (lymphocytes), filters blood, searches for body invaders, stores blood cells and destroys and disposes of old blood cells that no longer function due to age. So the spleen is part farmer, cleaner, detective, security guard, banker, hit man and undertaker. The spleen is located underneath the diaphragm on the left side of the abdomen near the stomach. It is shaped rather like a boxing glove.

The spleen takes blood from the heart's aorta — the largest artery in the body, where the blood travels through a vast channel of tiny blood vessels on the way to the liver. These tiny vessels are surrounded by large groups of lymphocytes. This blood movement through the spleen is "observed" by immune system T-cells as they look for any invaders of the body. If an invader is detected, a lymphocyte response is immediately triggered. Antibodies matching the invader are rapidly produced to overwhelm it. The spleen's blood vessels are also lined with macrophages, which consume and digest any debris in the blood, including the remains of any invader.

When the spleen is overworked it can swell and even rupture. This can happen as a result of diseases such as infectious mononucleosis (glandular fever). The spleen can also be ruptured by a serious injury to the abdomen.

CAN I LIVE WITHOUT A SPLEEN?

A seriously damaged spleen is no help to the body. If it can't be surgically repaired (splenorrhaphy) it may have to be removed (splenectomy). Although you can live without a spleen, having 1 is far better. Splenectomised individuals often have problems overcoming serious bacterial infections of the blood (sepsis). The spleen is the boxing glove to fight back.

WHAT IS A CHIMERA?
(Asked by C. Stanhope of Lafayette, California, USA)

In Greek mythology, a chimera was a fire-breathing monster with the head of a lion, the body of a goat and the tail of a serpent. A pretty mixed-up creature! In biology, a chimera is any animal, including a human, that has more than 1 type of genetically distinct cells. An example of a chimera would be someone with type O blood *and* type B blood. Normally, you'd have 1 or the other, not both. Chimeras gain their uniqueness from being the product of more than one zygote (a single diploid cell resulting from the fusion of male sperm and female ovum at fertilisation).

The stories of human chimeras are often fascinating. In Britain, a boy was recently reported in the medical literature as having the DNA in his skin cells indicating male gender (XY chromosomes). However, the DNA in his blood cells indicated female gender (XX chromosomes). In other words, his skin says he's male, while his blood says he's female. He was outwardly perfectly normal. One theory as to how this happened is that his mother's unfertilised egg somehow spontaneously divided before fertilisation, so that the egg was only partially fertilised (partial parthenogenesis). (Yes, a woman can just be "a little bit pregnant" in at least this sense.) The sperm got to just 1 of the two cells derived from the egg. The embryo then continued to develop but was part "all-mother" and part "mother–father".

In a US human chimera case also recently reported in the medical literature, a woman was discovered to have male DNA among her blood cells 27 years after giving birth to a male child. Investigators think that somehow "descendants of foetal cells 'escaped' during her pregnancy and persist in her system". Thus, the mother became a biological "blend of herself and her child". Strange.[19, 20]

WHAT DOES THE JUGULAR VEIN JUGGLE?

(Asked by B. Williams of the Bronx, New York, USA)

The jugular vein doesn't juggle anything. It's "jugular", not "juggler". The word jugular refers to the throat or neck. It is from the Latin word *jugulum* meaning "throat" or "collarbone", which derives from *jungere*, meaning "to join". The jugular vein is also known as the *vena jugularis*. It is actually not just 1 vein but 5, all with the same name. All 5 jugular veins move blood from various areas within the head, brain, face and neck down to the heart. The 2 largest of these are the right interior jugular vein and the left interior jugular vein. These carry huge volumes of blood, and are the 2 veins most often referred to as "the jugular vein". The other 3 are the external jugular vein, the posterior jugular vein and the anterior jugular vein. These are all smaller veins. When someone is experiencing a heart attack or congestive heart failure, the height and pulsation activity of the jugular veins are monitored in order to estimate whether or not the heart is keeping up or failing. From the standpoint of the body, the expression "going for the jugular", attacking a vital and vulnerable spot to show that you mean business, is certainly apt. If the jugular veins are severed, particularly either of the 2 interior jugular veins, the person can soon die from loss of blood and blood pressure. When someone attempts to cut someone's throat, they *do* mean business![21]

WHAT'S VICTORIA PRINCIPAL'S BLOOD DISEASE?

The actress Victoria Principal has recently disclosed that she has suffered for years from a rare and serious blood disease called disseminated intravascular hypercoagulation (DIH). DIH sufferers experience too much clotting of their blood. It can be thought of as the opposite of haemophilia, which involves too little clotting of the blood. DIH literally means too much clotting in the vessels throughout the body. "Disseminated" is from the Latin *disseminare* meaning "to sow" (as in "to widely spread seeds"); "intravascular" refers to inside the blood vessels; and "hypercoagulation" refers to too much clotting.

When you cut yourself and lose blood, the body has an automatic mechanism to stem blood loss. Blood proteins and platelets bind together and form clots in a process known as coagulation. Normal coagulation is important to health. But blood should not coagulate during normal movement through the vessels. Blood clots that are formed inside the body can be very dangerous. They can travel through the body and cause deep vein thrombosis (a blood clot in the arms, legs or pelvis), pulmonary embolus (a blood clot in the lungs), coronary thrombosis (a blood clot in the heart — aka a heart attack) or cerebral thrombosis (a blood clot in the brain — aka a stroke). Even if a heart attack or stroke is avoided, blood clots can still cause severe limb pain, difficulty in walking and even the need to amputate a limb.

All blood circulates through the heart. Abnormal blood clots usually form inside veins (blood vessels that carry blood to the heart) rather than inside arteries (blood vessels that carry blood away from the heart). DIH increases the risk of blood clots in the arteries as well as the veins. According to Dr B.H. Gobel, "in its acute form [DIH] may be life-threatening [and] may lead to irreversible morbidity and mortality".[22] The good news is it's treatable. Let's keep our fingers crossed for Victoria.

IS THERE ANY PART OF THE BODY THAT IS INDEPENDENT OF THE CIRCULATORY SYSTEM?

All parts of the body need to be connected to the circulatory system. All body systems are interdependent with the circulatory system, since all of them are supplied with oxygen, foods and other nutrients and have their waste products from metabolism removed. In addition to doing the tasks mentioned above, the circulation system distributes heat, hormones and other chemical messengers, helps fight diseases and infections, aids in the healing process and performs many other important jobs. All systems are connected through many kilometres of blood vessels and capillaries. These microscopic, thin-walled tubes allow for the exchange of materials between the circulatory system and every cell in our body. The lymphatic system helps in some of these functions as does interstitial fluid, a filtrate of blood. All systems are well served by the circulatory system.

HOW EXACTLY DOES THE BRAIN CONTROL BREATHING?

We don't know everything about every aspect of this process. Breathing is an automatic rhythmic action that persists without conscious effort whether we are awake or asleep. How this is done is a question that has intrigued many scientists for well over 100 years. According to Dr Patrice Guyenet,[23] specialised nerve cells (central chemoreceptors) are required to power the respiration rhythm. But the precise location and cell types involved are still not known for sure. Dr Guyenet proposes that these chemoreceptors are located in the brain stem and loaded with a transmitter substance called glutamate. Being in the brain stem, they can sense and react to changes in the pH in the cerebrospinal fluid, which signal the need for changes in breathing. However, according to Dr George Richerson,[24] the central chemoreceptors are found close to the midline blood vessels of the brain stem, allowing them to "taste" the pH of the blood. He also believes that the cells do not contain glutamate but serotonin.

ARE THE LUNGS LIKE BALLOONS?

(Asked by C. Hampton of Scarsdale, New York, USA)

It is a common misconception that the lungs are 2 empty sacs of air that, as we breathe in and out, inflate and deflate just like balloons. This notion may be easy to visualise and functionally descriptive, but physiologically it is incorrect. The lungs are soft, light and very elastic organs. The interior does not contain any empty air space. However, most of a lung is composed of a porous tissue much like a sponge. The main function of lungs is respiration. This exchange of oxygen for carbon dioxide takes place inside this porous tissue and within tiny air sacs call alveoli. Human lungs contain over 300 million of these tiny air sacs. Together they provide around 75 square metres (about 90 square yards) of surface area for breathing. The surface of a balloon holds the air inside. The surface of a lung lets air pass through.[15]

WHAT COLOUR ARE LUNGS?

(Asked by C. Hampton of Scarsdale, New York, USA)

The lungs of an adult are not pink or red as is generally believed. Only the lungs of a newborn infant are pink.

Adult lungs are usually white or a pinkish-grey. By adulthood, the lungs are spotted with dark grey or bluish patches. This discolouration is the inevitable consequence of breathing in carbon, dust particles, dirt and anything else that is part of normal polluted air. People would never start smoking if they could see what the smoke leaves behind in their lungs.[15]

ARE BOTH LUNGS THE SAME SIZE?

(Asked by C. Hampton of Scarsdale, New York, USA)

Both lungs are different in size. Generally speaking, externally humans and all other vertebrates have bilateral symmetry. This means that the left and the right sides of the body are virtually identical on the outside. However, humans and all other vertebrates are not symmetrical internally. This is genetically determined and starts from the body's earliest development as an embryo. The result is that there are differences between the internal organs on the left and right sides of the body. In humans, the heart, stomach and spleen are on the left, while the appendix and gall bladder are on the right. Likewise, there is less liver on the left side in order to make room for the stomach. And the lung on the left side is smaller in order to make room for the heart. The reason for the asymmetry of these organs has to do with the limited space available in the torso. For example, the intestinal tract is too long to remain in a straight line, so it must be looped and curled to fit into the abdominal cavity, while still leaving room for the other organs. The body doesn't waste space.[15]

CAN HUMANS LIVE WITH 1 LUNG?

Yes. We can and often do. There are people with lung cancer who have to have 1 of their lungs removed and they survive very well with only 1 lung.

DO OUR LUNGS GET BIGGER AS WE GROW?

Yes. All you have to do is look at a baby's chest and compare it with the chest of an adult and you'll see that the adult's chest is much bigger. Most of the chest (thoracic cage) is occupied by the lungs.

Lung size also varies in adults. Think of a tiny man and compare his chest volume with that of a large football player, for instance. The football player will have larger lungs than the short man.

IF AIR PRESSURE IS SO STRONG, WHY DOESN'T IT HURT US?

(Asked by Gillian Lacey of Newcastle, New South Wales)

Air exerts pressure equally on all sides of the human body. However, the body consists mostly of water. In fact, you could think of the human body as a combination of water and oil with some proteins thrown in for good measure, to give the whole thing a shape and to provide for some interesting behaviours. Water and oil are essentially non-compressible. They occupy the same volume no matter how much pressure you place on them. They might change state and become more solid, yet they would still occupy relatively the same volume. The pressure that the body exerts in an outward direction to counteract the air pressure is due to the molecular nature of the water and oil that make up our bodies.

A good example of this is a balloon. If you take an air-filled balloon down to the bottom of the ocean it ends up bursting and looking all shrivelled up, since the water pressure crushes the air-filled sphere. This is because the air within the balloon is compressible. If you completely fill the balloon with water then take it down to the bottom of the ocean, it still may remain un-popped and look like a balloon because the water is relatively non-compressible.

But there are 2 additional problems indirectly created by air pressure. First, we have air-filled lungs. However, as long as our lungs are open to the outside air via our mouth and nose, the air pressure is the same inside the lungs as outside. Second, we have air-filled ears, though only our middle ear is filled with air. You can feel the direct effects of this when you fly in a plane and your ears begin to hurt. The air pressure in the passenger cabin of a plane is maintained at about the air pressure of 1524 metres (5000 feet) above sea level. However, the air pressure in your middle ear is that found at sea level. Fortunately, the middle ear is directly connected to the outside by the eustachian tube, which opens in the back of the mouth. The mouth end of the eustachian tube is normally closed. Whenever we swallow, the eustachian tube opens and the air pressure inside our middle ear becomes equal to the pressure outside. This is why it helps to swallow as much as you can when you fly.

WHAT IS A COUGH? HOW AND WHY DO PEOPLE COUGH?

A cough is a vital mechanism for protecting the lungs. Coughing is a forceful, sharp sending out (expiration) of air resulting from contraction of the expiratory muscles of the lung against a closed glottis after a build-up of pressure in the chest. This results in expulsion of air at a high velocity which sweeps the airway free of debris. The glottis is the vocal apparatus of the larynx (voice box) and consists of the vocal cords and the opening between them.

A cough is triggered in 2 ways. The first is a mechanical stimulation of the larynx that causes an immediate expiratory effort in an attempt to clear material, e.g., a foreign body such as a speck of dirt, from the airway. The second is an enhanced mucociliary clearance in people with or without lung disease. Mucociliary refers to the mucus membrane and the hairs (including nose hairs) of the respiratory system. A cough may be stimulated by something internal (e.g., secretions or airway inflammation) or something external such as an irritant (e.g., smoke, dust or inhaled chemicals). A chronic cough is defined as a cough lasting for more than 3 weeks. The importance of a cough is most clearly seen in conditions associated with a weakened cough, such as in neuromuscular disease. The cough pathway involves cough receptors, sensory nerves, the vagus nerve, the cough centre in the brain stem, afferent nerves and effector muscles. An afferent nerve transmits impulses from the tissues to the brain and spinal cord. An effector muscle regulates nerve pathways by increasing or decreasing the pathways' reaction rate.[25]

WHY DO I GET OUT OF BREATH SO EASILY AT A HIGH ELEVATION?

This is simply a matter of air pressure. As you rise in altitude, the air pressure decreases. At the same time, the amount of oxygen in the air decreases proportionately, remaining at a little less than 21 per cent of the total air pressure. Both oxygen and carbon dioxide influence respiration. Carbon dioxide controls respiration primarily by acting directly on respiratory centres in the brain. Oxygen controls respiration by acting on the chemical receptors located in the carotid and aortic areas of the lungs and heart. These chemical receptors transmit signals to the brain to stimulate respiration. Under normal conditions, oxygen control of respiration is almost insignificant compared with that of carbon dioxide. It seems strange that oxygen would play such a small role in the regulation of respiration, but there it is. The reason for this is that the respiratory system is capable of supplying the blood with enough oxygen even when respiration has been slowed significantly. Respiration can drop tremendously without significantly affecting oxygen in the tissues. On the other hand, small changes in respiration have a tremendous effect on the amount of carbon dioxide in the tissues.

HOW CAN TIBETAN VILLAGERS LIVE AT SUCH HIGH ALTITUDES?

Over the past few years, researchers have discovered that Tibetan villagers living at high altitudes (over 4000 metres, or 13,000 feet) are genetically adapted to doing so. Within the past 10,000 years or so, these mountain-dwelling Tibetans have evolved genetically to enable them to live in their environment. Compared with other humans, these Tibetans have at least 1 extra gene that helps their blood cells bind more oxygen. This also appears to boost their reproductive fitness under the low-oxygen conditions experienced at very high elevations. They can breathe better and they can breed better than could other humans experiencing the identical mountainous conditions. As Tibetans evolved over thousands of years, many who did not have the "high altitude/high oxygen" gene must have perished. Those who did have the gene were more likely to survive and thus successfully pass it on to their offspring. Over the generations, as such Tibetans increased in proportion if not in number within the population, nearly everybody had the "high altitude/high oxygen" gene.

This evolutionary process is hardly surprising. Over the past 30 years, nearly 1500 people have successfully scaled Mount Everest. Some used extra oxygen in tanks, whilst others relied only on that available freely in the air. Many died during the climb up or during the descent. But it is fascinating that the chances of survival were nearly 3 times as high for those who used extra oxygen in tanks. In 2004, Dr Hans Hoppeler, a sports physician from the University of Bern in Switzerland, found that mountain climbers can experience cell damage to their muscles as a result of the oxygen limitations at high altitudes. So it's easy to imagine why the Tibetans may have needed to evolve.[26]

WHAT CAUSES SOMEONE TO HAVE THEIR BREATH "KNOCKED OUT"?

(Asked by John Miller of Sunnyvale, California, USA)

For an adult of average size, the lungs can hold a total of about 5 litres of air. During normal breathing, the tidal volume of air entering and leaving the lungs is about ⅒ of the total volume, or 500 millilitres. Through conscious, deep breathing, a person can inhale a full 3 litres of air. This is called the inspiratory reserve. They can exhale an extra 1 litre. This is called the expiratory reserve. This adds up to 4 litres of air being moved during deep breathing, but the total volume of air in the lungs is 5 litres. The remaining 1 litre of air in the lungs is called the residual volume. The residual volume is the result of the design of the lungs. This design prevents them from collapsing completely like a balloon, which would allow them to expel all of their contents.

Breathing is controlled both consciously and unconsciously by the brain through a complex system of sensory and motor nerves. The motor nerves run to the diaphragm and the intercostal muscles located between the ribs, allowing the expansion and contraction of the chest during breathing. The sensory nerves run to various receptors that tell the brain how much oxygen is in the blood, how much carbon dioxide is in the blood, how much air pressure is in the lungs and how much tension is being placed on the chest muscles during breathing. These last 2 give important feedback to the brain to allow it to coordinate the muscles so that breathing runs smoothly. When someone or something impacts against your chest with enough force, the air forced out of the lungs can exceed the normal expiratory reserve, such that the residual volume is below the normal limits of the lungs. When this happens, miscues from the sensory nerves confuse the brain, so that it cannot "feel" the lungs to take a breath, until the natural elasticity of the lungs recoils from the blow and returns the volume to above normal residual levels. This recoil takes at most a couple of seconds, during which time the individual is unable to breathe. It can get pretty scary. So the phrase "having your breath knocked out" is actually very accurate.

WILL WE EVER DEVELOP AN ARTIFICIAL HEART?

Reviewing research in the development of an artificial heart, Dr T.S. Guy writes that "After years of hope and disappointment, artificial heart technology may be on the verge of efficacy."[27] A partial artificial heart is already available. It is called a left ventricular assist device (LVAD). The LVAD temporarily takes over much of the heart's pumping action in patients awaiting heart transplants. The entire device can be implanted, except for the battery pack, controller and cord. It is possible to implant a LVAD permanently in patients who would otherwise require a transplant. Currently, only about 1 in every 15 people who needs a heart transplant is able to get one.

About 40,000 Australians have an artificial cardiac pacemaker implanted when the heart's natural pacemaker fails — causing the heart to beat erratically and dangerously. This matchbox-sized device consists of a generator, containing a battery and electrical circuitry, and 1 or 2 insulated wires. It sends out regular electrical impulses that stimulate the heart muscle and regulate the heartbeat. Artificial heart valves that control blood flow between the heart's chambers are now available. These are made of metal and synthetic materials or can be fashioned from animal tissue (usually a pig's heart valve).[28-30]

WILL WE EVER DEVELOP ARTIFICIAL BLOOD VESSELS?

Angiogenesis is the process of growing new blood vessels. According to Dr Judah Folkman from the Department of Cardiology at the Harvard Medical School, specific molecules can inhibit blood vessel growth and thus create "smart" blood vessels. Such "smart" blood vessels can then be "trained" to "starve" cancerous tumours of their blood supply and thus kill the tumour. "Smart" blood vessels can also be "trained" to boost the healing rate of any injured or damaged organ. Dacron blood-vessel grafts are used to bypass diseased portions of arteries (for example, in the leg) or to replace the artery entirely.[28-30]

WILL WE EVER DEVELOP AN ARTIFICIAL LUNG?

Experiments with an artificial lung are taking place at a number of medical centres around the world. One version is being developed at the University of Pittsburgh. About the length of a loaf of bread, it consists of about 1000 tiny, porous, hollow-membrane fibres surrounding an elongated central balloon. Inserted into the patient's chest cavity and right atrium, it is immersed in oxygen-depleted blood. Oxygen pumped into an external tube flows into the fibres and then diffuses through tiny pores into the blood. Carbon dioxide travels in reverse. It goes from the blood into the fibres and then exits the body through a second tube.[28-30]

The human heart beats 70 times per minute. The hummingbird's beats 1300 times per minute. The heart of a blue whale beats 10 times per minute.

For reasons unknown, female patients reject heart transplants more often than men do.

Leeches make you bleed because their saliva contains a substance that keeps your blood from clotting.

Your heart will thump approximately 42,075,900 beats per year — give or take a few thousand.

During a 24-hour period, the average human will breathe 23,040 times, exercise 7 million brain cells and speak 4800 words.

If laid out in a straight line, the average adult's circulatory system would be more than 100,000 kilometres (62,000 miles) long — enough to circle the Earth 2 and a half times.

CHAPTER 11

THE STOMACH & INTESTINES

Food enters the stomach. It gets crushed and reduced in size by chewing (mastication) and gets saturated with saliva. The stomach provides the basic functions that assist in the early stages of digestion and prepares food for further processing in the small intestine.

The stomach serves as a short-term storage reservoir that allows a rather large meal to be consumed quickly and dealt with over an extended period of time. It is the place where substantial chemical and enzymatic digestion is initiated, particularly of proteins.

Vigorous contractions of gastric muscle mix and grind foodstuffs with gastric secretions, resulting in liquefaction of food. Food, in the now liquefied form, is then delivered to the small and large intestines, where the digestion process is completed — and we live to eat another day.

WHY DO I GET HUNGRY?

(Asked by T. Banfield of Kitchener, Ontario, Canada)

Hunger is defined scientifically as an uneasy sensation this is normally occasioned by a want or craving for food. It is a myth that the sensation of hunger has mostly to do with an empty stomach. The truth is more complicated. Hunger is dependent on the direct stimulation of sensory nerves in the stomach and intestines and indirect sensations from other organs when they are exhausted from lack of nutrients. It begins when certain nutrients are missing in the blood. A message is then sent to the hunger centre of the brain, which is located in the hypothalamus. From there, the brain activates the stomach and intestines. Growling and rumbling (borborygmus) occur as a consequence of rhythmic contractions. These combine with the all-important psychological reaction to all of these stimuli.

Hunger stops when we have sufficient nutrients in the blood. This causes the brain to inhibit stomach and intestinal activity. However, merely having sufficient nutrients in our bodies is not always enough to stop us eating. The sensory cues in the mind from the tastes, smells, sight and even touch of food, plus the memory of our delight in eating, combine to trigger a psychological response that may drive us to eat food even when we don't "need" it.[1-2]

WHY DO I SOMETIMES GET A HEADACHE WHEN I'M HUNGRY?

(Asked by T. Banfield of Kitchener, Ontario, Canada)

The sensation of hunger also involves serotonin. Serotonin is a substance used in sending and receiving signals in the brain (neurotransmitter). It is also involved in thirst, mood, sleep and several other body processes. A lack of serotonin is one of the causes of migraine headaches. When the sensation of hunger occurs, a person with a sensitive serotonin system can experience a headache.[1-2]

WHAT IS DIGESTION?

Chew on this one! Contrary to common belief, food is not digested in the stomach. In fact, the stomach plays only a relatively minor role in the process of digestion. The stomach's primary purpose is simply to store ingested food and prepare it for the digestive process. Digestion works like this: after food is chewed (masticated) and mixed with saliva, it is known as bolus. The bolus enters the stomach, where hydrochloric acid, pepsin and other enzymes are mixed with it for 3 to 4 hours in preparation for digestion. This mixture is known as chyme (pronounced "kime", to rhyme with "time"). Chyme is also the scientific name for what comes up when we vomit. The storing and mixing of chyme is the primary purpose of the stomach.

Digestion actually takes place in the small intestines. When the chyme has been properly prepared, a small valve at the bottom of the stomach opens. This allows the contents to enter the intestines. With the addition of a combination of bile from the liver, juices from the pancreas and other intestinal fluids, the absorption of the chyme into the intestinal wall takes place. When the large intestine removes water from the undigested residue the process of digestion is completed. What is left is eliminated from the body. The whole process takes 20 hours, more or less.

WHAT KEEPS STOMACH ACID FROM EATING MY STOMACH?

Yucky as it may seem, we owe the protection of our stomach to mucus. Without mucus, stomach acid (which is strong enough to dissolve metal) would eat through the stomach wall and keep right on going. The stomach wall is made of many layers. The outermost layer (serosa) is thin and tough. Inside the serosa are three layers of involuntary muscle that mechanically grind and mix swallowed food. The innermost layer (epithelium) consists of many specialised cells. One type, the parietal cells, secrete hydrochloric acid (HCl). The parietal cells, along with other cells, also secrete enzymes that dissolve proteins. However, the epithelium also has many mucus-secreting cells. The mucus creates a protective covering for the epithelium that buffers and neutralises the stomach acid, thus protecting it from the acid's fury. There's another reason why the enzymes do not eat the stomach wall: they act at a low pH when they enter the stomach channel (lumen). The pH near the surface of the epithelium is much higher. This is due to the overlying mucus that shields the epithelium from the acid. The result of this biochemical buffering of enzymes through the stomach is that the enzymes do not become active until they have safely gone beyond living tissues.

IS STOMACH ACID CAUSTIC ENOUGH TO BURN THE SKIN?

(Asked by Chris Bernard of Wentworthville, New South Wales)

The acidity of the stomach normally is in the pH range of 1.5 to 3.5. This is a very concentrated acid. If it were put on your skin it would cause a great deal of irritation and burning. An equivalent effect happens during heartburn, when acid from the stomach refluxes (i.e., flows backward) into the oesophagus and burns the lining of the oesophagus. Is stomach acid caustic enough to burn the skin? You better believe it is.

WHAT CAUSES A STOMACH ULCER?

A stomach ulcer is more correctly called a gastric ulcer. When it occurs in the duodenum (a short segment of the small intestine that connects directly to the stomach), it's called a duodenal ulcer. Both gastric and duodenal ulcers are also known by the broader term peptic ulcers.

If a peptic ulcer is left untreated, it can lead to serious complications such as severe bleeding, perforation of the stomach or duodenum lining, and peritonitis. Peptic ulcers are usually raw patches 1 to 2 cm in diameter. When viewed with a gastroscope, they look rather like mouth ulcers. Gastric ulcers are rare in people before middle age. Duodenal ulcers are very common. They occur in 10 per cent of people at some point in their lives. Peptic ulcers may be acute (symptoms appear and improve quite quickly) or chronic (symptoms go on for a long time). Acute peptic ulcers often happen in groups of more than 1; these may produce no symptoms and often heal without any long-term consequences. Chronic peptic ulcers are deeper, usually occur individually, cause symptoms and leave a scar when they heal.

We used to think that peptic ulcers were caused by psychological factors such as stress and mental imbalance. However, we now know that a microbe called *Helicobacter pylori* (*H. pylori*) causes many peptic ulcers. The organism lives in the stomach's mucus layer and "does its thing" by causing long-term irritation of the epithelial lining of the stomach or duodenum. Peptic ulcers can cause symptoms of indigestion and severe pain. A peptic ulcer can also result when the stomach produces excess acid or there is insufficient mucus to protect the lining from damage. There are several factors that increase a person's risk of getting a peptic ulcer: (1) infection with *H. pylori* — this microbe is almost always present in people with peptic ulcers, but is also found in people without them; (2) regularly taking certain medicines, particularly aspirin and other non-steroidal anti-inflammatory drugs; (3) smoking; (4) drinking alcohol in excess; (5) stomach cancer. See your family physician if you have any abdominal pain.[3]

WHAT DOES THE SMALL INTESTINE DO?

(Asked by Chris Bernard of Wentworthville, New South Wales)

The small intestine is a hose-like muscular portion of the digestive system extending from the lower end of the stomach to the large intestine. The fairly narrow small intestine is a tube-like structure that winds back and forth.

The contraction of its muscular walls propels food onward while digestion is completed. The way the walls contract as they move food along is rather like how a snake digests a meal. This process in humans is called peristalsis. *Stalsis* is a Greek word meaning "contraction". Lots of small projections called villi in the intestinal lining absorb the digested food for distribution by the blood to the rest of the body. The small intestine joins the large intestine (colon), where most of the water content of the remaining mass is absorbed by the body.[4]

HOW DOES A TAPEWORM SURVIVE INSIDE MY BODY?

Surprisingly, this is an often-asked OBQ even though tapeworm infestation is less of a problem in modern industrial societies, since the quality of our food is higher than in earlier times. Thank the inventor of the refrigerator for much of this.

Tapeworms come into the body via contaminated food. Many organisms live on and in the human body. A tapeworm can easily survive and thrive indefinitely inside us. A tapeworm in a human can range in length from 5 millimetres (⅟₂₅₀ of an inch) to an incredible 15 metres (50 feet)! Tapeworms have no digestive tract so they must eat food already digested by another animal. That is precisely what they do as a parasite inside our intestines. They absorb nutrients directly through their skin (cuticle). They also reproduce inside us. Tapeworms are simple in design, but ruthless in action. They consist of 2 organs. The first organ (scolex) anchors the beastie to the wall of the intestine with suckers and hooks. The second organ (proglottid) is really a series of organs that grow out from the scolex, with each having full reproductive capability. Proglottids form a chain of varying length. The last segment of the chain eventually breaks off and is passed out with faeces.

Tapeworms resist being destroyed by the body's immune system or digestive juices. They cause health problems around the world and can even kill, since they rob us of nutrients, block our intestines, take up space in organs and stop them from functioning normally. A tapeworm cyst can settle in the brain, eye, liver and elsewhere. There are many species of tapeworms, and not all of them can infest humans. Although little is known about the origins of tapeworms in humans, it is well known that some tapeworms live in both animals and humans. Some tapeworms have such a complex life that they are required to live first in a herbivore (such as a cow) and then in a carnivore (such as a human), and only then can they reproduce.[5]

WHY DOES MY URINE STREAM SOMETIMES SPLIT IN 2?

(Asked by C. Barrie of New York, New York, USA)

There are at least 4 possible explanations for this quirky body behaviour among males. It does not seem to occur among females.

Two streams may result when a male lies on his side with his legs closed for a long period of time. This could squeeze the urethral orifice and the urethra itself. If he gets up quickly to urinate, the urine may be partially blocked for a time and may form a split stream for a few seconds.

Two streams may result when the urethral orifice is partially closed due to exposure to cold, such as when one has been swimming in cold water and then attempts to urinate. Males often observe that swimming in cold water causes their penis and testicles to temporarily shrink. This is due to the movement of blood to warm the extremities.

☞ Two streams may result from the presence of dried body fluids in the end of the urethral orifice. This may cause the sides of the orifice to stick together so that when the force of urine strikes it the stream initially opens 2 separate breaks in what was a plug. By the time the bladder is empty, the 2 streams will have united to form 1.

☞ Two streams may also be caused by an unusual (but not unknown) anatomical anomaly in which there are actually two openings in the urethra.

WHAT IS PRIAPISM?

Priapism is the persistent abnormal erection of the penis. The erection can last for hours. It can be a medical emergency. Men can die from it. There is no sexual desire with this form of erection and the sufferer is often in great pain. Priapism is caused by disease or injury to the penis, but also to the spinal cord, bladder or kidneys. It is a problem of the corpora cavernosa (CC) of the penis. The CC are 2 chambers running the length of the penis filled with spongy tissue. A man has an erection when the tissue fills with blood. Priapism occurs when the blood remains in the CC for too long for whatever reason.

There are 2 types of priapism: arterial high-flow and veno-occlusive. Arterial high-flow priapism usually occurs after a rupture of the cavernous artery of the penis. There is a temporary heavy blood flow to the CC. This is usually the result of a less serious injury to the CC, and there is less risk of permanent damage to the CC and less pain. Ironically, the injection of medications into the penis to treat impotence can sometimes cause this type of priapism. Veno-occlusive priapism usually occurs after complete CC tissue blockage. This is usually the result of a more serious injury or disease. There is more risk of permanent damage to the CC and more pain. This type of priapism can result in fibrosis (degeneration of the tissues involved) and eventually to a loss of the ability to achieve an erection.

Priapism can occur in males of any age, beginning from infancy.

In 1 study, 40 per cent of patients with sickle cell disease — an inherited disorder of the red blood cells — reported at least 1 episode of priapism, with the peak incidence occurring at about age 20.

Believe it or not, priapism can occur in women too. But according to Dr Martin J. Carey,[6] "Priapism is primarily a disease of males. Priapism of the clitoris has been described very rarely."

WHY DO HUMAN MALES HAVE 2 TESTICLES AND NOT JUST 1?

Why not 3 or even 4? Perhaps 5 or 6? The more the merrier? Men have 2 testes for producing sperm for the same reason as women have 2 ovaries for holding and releasing eggs for reproduction. All animals with a spinal cord have 2 gonads. Gonads are the sex organs (ovaries in females, testicles in males) that produce gametes (the specialised cells produced by cell division in sexual reproduction). When a human is still an embryo, the gonads develop from embryonic kidney tissues lying on either side of the spinal cord. Producing a vertebrate with only 1 gonad would require some major restructuring of the urinary tract, major rerouting of the circulation system towards the caudal (tail) region of the body, and major refashioning of the genital "apparatus". (It has become a fashion among physicians in polite circles these days to refer to the "apparatus" as one's "unit".) Any evolutionary or survival advantage of having 1 gonad instead of 2 would have to outweigh the disadvantage of losing a kidney and half a bladder, and the rearranging of our body structure, which nature has been perfecting in humans for at least 3 million years. With only 1 gonad and without such rearranging, the human body would be quite out of balance.

Another reason for having a second gonad is that having a "spare" has its advantages. If 1 is damaged, the other still gets the job done. That's why we carry a spare tyre in our cars. Why not 3 or more gonads? Or in the case of males, why not 3 or more testicles? Having 2 or 3 spares would be a needless luxury and would take up more space in the body than it was worth. Nature is pretty smart when it comes to providing just the right space allocation in the body to balance everything in order for the species to survive and multiply. After all, who carries more than 1 spare tyre in their car?[7]

WILL SITTING ON COLD CONCRETE GIVE ME DIARRHOEA OR HAEMORRHOIDS?

No. It will just feel uncomfortable. It is a myth that sitting on cold concrete or even a block of ice will give you diarrhoea or haemorrhoids (piles). Diarrhoea is a condition involving loose and watery stools that occur more than 3 times in 1 day. There are 7 ways that diarrhoea can be caused — sitting on something cold is not 1 of them. Diarrhoea may occur through: (1) a bacterial infection, e.g. from drinking contaminated water; (2) a viral infection; (3) a food intolerance; (4) an intestinal parasite; (5) a reaction to medicines; (6) an intestinal disease such as celiac disease or inflammatory bowel syndrome; or (7) a bowel disorder such as irritable bowel syndrome. A haemorrhoid is a varicose dilation of a vein in the superior or inferior haemorrhoidal plexus area of the anus. A haemorrhoid results from stress on the vein (venous pressure), not cold.

WHAT IS THE PANCREAS AND WHAT DOES IT DO?

(Asked by Sarah Burgess of One Tree Hill, South Australia)

The pancreas is an elongated gland located behind the stomach. It is involved in the body's digestion and usage of fats, carbohydrates, proteins and acids. The main job of the pancreas is to achieve the exact biochemical mix necessary for successful digestion. It is one of the many wonders of the human body. What the pancreas does is a bit complicated. It releases digestive enzymes and bicarbonate ions into the pancreatic duct, then interacts with the common bile duct and empties its contents into the duodenum. It is in this area that any food that was beginning to be digested can be broken down into its elementary parts and absorbed by the body.

The pancreas secretes enzymes that include trypsin, chymotrypsin, carboxypeptidase and elastase. These enzymes break the peptide bonds of protein. The pancreas also produces lipase, which breaks up triglycerides into individual fatty acids. And the pancreas secretes amylase, which splits polysaccharides into glucose and maltose. Plus, it produces hormones involved in organic metabolism and blood glucose levels. These include insulin, glucagon, somatostatin and pancreatic polypeptide. The pancreas also produces ribonuclease and deoxyribonuclease, which split nucleic acid molecules. It provides the bicarbonate ions that lower the acidity of the digestive tract in the duodenum by neutralising much of the acidity of material arriving from the stomach. The stomach is able to withstand a highly acidic environment due to its special protective mucus lining, but the small intestine does not have such a lining and is thus unable to protect itself. So it is very important that there be a decrease in the acidity of material entering through the duodenum. The pancreas saves the day by performing this role too. It truly does a remarkable job!

WHAT IS BIGOREXIA?

Bigorexia nervosa is a new eating disorder that is the opposite of anorexia nervosa. In bigorexia, one thinks that one's body must become *bigger*. As a result, one will exercise, lift weights, eat enormous amounts of food and often take steroids. Body size is a 24-hour-a-day preoccupation. The sufferer's behaviour is obsessive, compulsive and delusional. For example, even though someone may be tall and muscle-bound, they believe they are short and weak. Some are so ashamed of their bodies that they refuse to leave their homes. They drop out of school and lose jobs. Some become depressed to the point of being suicidal. According to Dr Eric Hollander,[8] at least 20,000 males in the US suffer from biogorexia. It affects mostly men. In this too it is the opposite of anorexia. According to Dr Harrison Pope, a professor of psychiatry at Harvard Medical School, in 1 study, 10 per cent of bodybuilders showed some signs of bigorexia. It is also known as muscle dysmorphia. It is so new that it has not been recognised as yet as a true psychiatric illness.[9]

WHAT IS ORTHOREXIA?

Orthorexia nervosa is another new eating disorder. It refers to the pathological obsession with biologically pure and immaculate food. Orthorexics exclude from their diet any food that may be the slightest bit contaminated through exposure to impurities. "Impurities" may include dust, dirt and artificial additives and preservatives, as well as pesticides. How food is prepared is often obsessed over by the orthorexic. It is very difficult to get their diet right and this becomes a 24-hour-a-day obsession. The term orthorexia comes from the Greek word *orthos* (straight, proper) and *orexia* (appetite). There have been only 6 entries in the medical literature for orthorexia. Like bigorexia, it is so new that it has not been recognised as yet as a true psychiatric illness. In the latest study, Dr M.L. Catalina Zamora and 3 other Spanish doctors report on a patient who "responds to the characteristics of orthorexia".[10, 11]

DOES IT HARM YOUR BOWELS IF YOU SUPPRESS A FART?

(Asked by E. Perrins of Eastwood, New South Wales)

Flatus is the medical term for breaking wind, passing gas, or the many other euphemisms we attribute to this body activity. *Flare* is Latin for "to blow". *Flatus* means "breath", "puff" or "wind". It is very descriptive. In physiology it refers to the wind or gas generated in the stomach or other cavities of the body.

There are differences of opinion as to whether or not suppressing flatus can harm you. It must be said that suppression is only a temporary measure. When you're about to fall asleep your muscles relax and flatus will "slip out". So you can't hold it in forever. People have believed for centuries that retaining flatus is bad for health. The Roman emperor Claudius passed a law that made flatus fashionable at banquets. This was out of concern for the damage that suppressing flatus could do to his guests. Benjamin Franklin wrote a book called *Fart Proudly*, which recommended allowing nature to take its course whatever the social occasion. There was a widespread belief that a person could be poisoned or catch a disease by retaining flatus.

Today, physicians usually say that there is no particular harm in holding in flatus. Flatus will not poison you, hurt your stomach or damage your bowels. It is a natural component of your intestinal contents. Probably the worst thing that can happen is a stomach pain from withholding a normal body process for too long. However, one doctor is on record as saying that it is theoretically possible to cause haemorrhoids by withholding flatus for too long. But most doctors say that's just a lot of hot air.[11]

WHERE DOES A FART COME FROM?

(Asked by E. Perrins of Eastwood, New South Wales)

The gas in our intestines comes from a number of sources. These include the air we swallow when we breathe and when we eat. This is particularly the case if one is a mouth-breather or eats too fast. Gas is also produced by chemical reactions in our gastrointestinal (GI) tract and by normal bacteria living in our GI tract. Gas seeps into our intestines from our blood too.[11]

WHY DO UNRIPE APPLES MAKE ME SICK?

(Asked by Kuranda Seyit of Bayonne, New Jersey, USA)

Better to ask your physician about any question regarding your health rather than a humble anthropologist like me. However, there is a widespread myth here. Although anyone can have an allergy to just about anything — including apples — and that can make a person sick, it is not necessarily the case that unripe apples make people sick *per se*. According to the US Department of Agriculture, both ripe and unripe apples are digestible by the stomach, which does *not* distinguish between the 2. The USDA adds that as long as the apple is chewed sufficiently, a green (unripe) apple will not cause sickness.

Yet there is a difference between ripe and unripe fruit, including apples. An unripe apple contains more acid and more long-chain polysaccharides (pectins) with a high molecular weight than ripe apples. Pectins are present in the cell walls of all plant life and form a kind of cement that helps keep cell walls together. Digestion of too many of these pectins can cause the development of gas in the stomach. People differ in their ability to handle acids and gas in their digestive system. That means that some people can and do have problems eating unripe apples, especially when they eat too many of them at one time.

HOW LONG DOES IT TAKE TO DIGEST SOMETHING COMPLETELY?

(Asked by Stan Crowley of Wiley Park, New South Wales)

This question seems to come up again and again. What you put into your mouth has to travel through the entire gastrointestinal tract before coming out the other end. A mouthful of food (the bolus) takes approximately the following times to pass through the various regions of the digestive tract:

Oesophagus — 5 to 7 seconds
Stomach — 2 to 6 hours
Small intestine — 3 to 5 hours
Large intestine — 3 to 10 hours

This totals about 20 hours, more or less. Of course, the process could be faster if you suffer from diarrhoea, or slower if you suffer from constipation. And some types of food move through the digestive system faster than others. According to Dr David Ng,[12] "the time for food to travel from one end to the other probably ranges from 20 to 30 hours". Indeed, there is even a cultural component to this. Reputedly, some individuals in Italy normally have only 1 bowel movement per week. Yet it is the cultural practice of other southern Europeans to have at least 1 bowel movement per day.[13]

If you are prone to constipation, chances are your rectum will be larger than normal. Dr J. Klijn and 4 colleagues from the University Children's Hospital in Utrecht reported in 2004 on their study of 49 children. It was found that constipated children had an average rectal diameter of 4.9 cm compared with an average rectal diameter of 2.1 cm for non-constipated children. This enlargement can persist into adulthood.[14, 15]

WHAT IS VOMITING?

Vomiting is a very complex act controlled by a vomiting centre at the base of the brain. That's right, vomiting is really an act of the brain and not the stomach. The stomach is shaped rather like a boxing glove. When you vomit, it feels as if you've been hit in the stomach by the current heavyweight champion. Despite the awful sensation produced in the stomach when vomiting, the stomach actually plays virtually no role in the physical act of ejecting undigested matter through the mouth. And it's not just the contents of the stomach that are forced out during vomiting, but also much of the material from the upper small intestine.

Vomiting occurs in quick stages. The first stage is strong. It consists of a series of reverse contractions of the upper small intestine. Next, the lower portion of the stomach contracts. After that, these successive contractions force the partially digested material from the small intestine and lower stomach into the main body of the stomach. The digestive tract is emptied as its contents move from the beginning of the large intestine up 3 metres (10 feet) of small intestine, into the stomach, up through the throat and out the mouth. Surprisingly, during vomiting the stomach remains completely relaxed while abdominal muscles and the diaphragm push in unison to compress the stomach and squeeze the contents up and out. That's why abdominal muscles feel so sore after vomiting. When you vomit, the body does something else it very rarely does: respiration is involuntary stopped to prevent aspiration of the vomit into the lungs.

There are many reasons for vomiting. These include food poisoning, indigestion, tonsillitis, obstructions, infections, motion sickness, smells, brain tumours, meningitis, intestinal parasites and morning sickness during pregnancy.

IS BED-WETTING GENETIC?

(Asked by Anne Perry of Capetown, South Africa)

This is a very interesting question. Researchers claim they have isolated a defective gene responsible for bed-wetting — the childhood problem that aggravates parents and embarrasses children. An estimated 500,000 Australian children over the age of 6 wet their beds at least once a week. Researchers now believe that more than half of those cases are genetic in origin, rather than psychological. A research team reported initial findings in 1995 that pointed to the location of the defective gene on chromosome 13.[16, 17] Their findings were elaborated upon in another study published in 1998. This study more closely identified the defective gene as being located on chromosome 22. Both studies involved the analysis of information on 11 children and 26 grandchildren with a history of the disorder. The team has now been able to "map" the position of "the bed-wetting gene", which they call PNEG. PNEG stands for the "primary nocturnal enuresis gene". According to the head of the research team, Dr Hans Eiberg, there may be as many as 10 genes on the segment that they have identified. One of them, a brain chemical receptor called HTR₂? "appears to be a good candidate as the cause of bed-wetting". Dr Eiberg adds that "preliminary evidence suggests that children whose bed-wetting is genetic in origin are the most responsive to drugs designed to halt the problem. Our discovery should also provide reassurance to parents who fear that the bed-wetting may result from their own child-rearing practices, from emotional trauma in their children or from simple rebellion."[18] The discovery of the bed-wetting gene is likely to demystify the problem of bed-wetting generally, solve the source of the problem in many bed-wetting cases, and help remove much of the child's shame and parental concern about this long-standing child development problem.

Bed-wetting, which is given the medical term enuresis, is common in very young children. Approximately 20 per cent of 4-year-olds experience enuresis. Approximately 10 per cent of 6-year-olds experience enuresis once a week or more. Although a nuisance for parents at any age, it is not considered a problem until the child is about age 7. An estimated 12 per cent of boys and 6 per cent of girls at age 12 still experience enuresis. Children almost always stop wetting the bed sooner or later. However, for many children and their parents, sooner is definitely better than later. Bed-wetting can have a serious effect on a child's self-esteem and can often be a source of stress for their family.

Enuresis comes in 2 forms: primary and secondary. PNEG relates only to primary enuresis. Primary enuresis characterises about 75 per cent of cases. It affects children who have never had a significant period of being dry at night. Secondary enuresis characterises the remaining 25 per cent. It affects those children who have been dry for at least 6 months and then revert to wetting the bed. Researchers now believe that most cases of primary enuresis are genetic in origin. They also believe that most cases of secondary enuresis are caused by emotional trauma. Such traumas could include the first day of school, being bullied, fears, family stresses and problems, even the birth of a sibling. Nevertheless, a medical problem, such as a bladder infection (which can be easily treated) or other conditions (which may or may not be easily treated), should never be ruled out in a case of secondary enuresis.

Child development research and expert opinion have long established that children do not wet the bed on purpose, but some parents still hold to this outdated notion. It is always a mistake to blame the child for bed-wetting. In fact, nearly 80 per cent of children who wet the bed have a parent who was a bed-wetter at the age of 5 or beyond. However, whatever the cause of bed-wetting in a particular child's case, emotional problems can exacerbate the issue if parents pour shame and guilt upon the child. In any case, a family physician should be the first source of treatment. If needed, he or she can make a referral to a specialist paediatrician or even a specialist paediatric urologist. Most children who wet the bed are perfectly healthy and can usually be treated without many laboratory tests. Several options for treatment are available and effective for children over the age of 6. No single method of treatment is ideal for every patient.

Doctors will usually rule out obvious causes of enuresis. Surprisingly, some parents will give their child a large drink just before bedtime. In these cases, it is no wonder that the child has a bed-wetting problem. These are the easy cases to solve. For more difficult cases, a doctor may recommend a behavioural approach first. This approach "trains" the bladder to hold more urine and usually includes some form of "bed-wetting alarm". It might start with simply rousing the sleeping child from bed at regular intervals during the night. According to many doctors, a reward system is often helpful in getting the child interested and involved in overcoming the difficulty. Next, a doctor may recommend medication, or a combination of medication and behavioural modification. Medications help the bladder hold more urine. There is no evidence that such medications have any harmful side effects, but many doctors believe it is important not to rely on medication alone to treat the child. Child development experts often note that parents should never be reluctant to discuss their child's bed-wetting experience with the family physician.[19]

WHAT IS THE APPENDIX USED FOR?

(Asked by Debbie Blair of Darwin, Northern Territory)

The full name of the appendix is the vermiform appendix. It's a worm-like appendage that projects off the cecum and is found in the lower right quadrant of the abdomen. *Vermis* is Latin for "worm". The cecum is a pouch connected to the ascending colon and constitutes the beginning of the large intestine. Cecum is from the Latin *caecus*, meaning "blind".

Many experts claim that the vermiform appendix is a vestigial organ that no longer serves any physiological function for the body and is merely a leftover from our evolutionary ancestry. However, many other experts dispute this. A vermiform appendix of some type or other can be found in nearly every mammal known to biology. This suggests that the appendix is important to mammalian physiology. Further evidence of the importance of this organ is found in the fact that it has its own blood supply, and an independent mesentery. (Mesentery is the double layer of peritoneum tissue that connects a part of the small intestine to the posterior wall of the abdomen.) This indicates that the appendix is not a part of some other organ. Nevertheless, the actual function of the vermiform appendix remains a mystery. According to Dr Alfred Romer and Dr Thomas Parsons, the major importance of vermiform appendix "would appear to be financial support of the surgical profession".[20] This is a rather tongue-in-cheek remark, yet it does emphasise the fact that humans live perfectly well without a vermiform appendix.

Current medical thinking points to the fact that the vermiform appendix contains a great number of lymphatic channels. Lymph fluid is the body's primary means of combating infection and foreign bacteria. The appendix also happens to feed into the cecum at the beginning of the large intestine. The large intestine is the home to many types of bacteria. *E. coli* is the most common of these. Not all of them are harmful and many may play an important role in our digestion. However, bacteria populations can get out of control and the lymphatics of the vermiform appendix may be important in this process. Therefore, the vermiform appendix may help to regulate the bacteria population in the body, but we won't die without it.

According to Dr Carol Shoshkes Reiss, "the appendix is not a specialised immune organ and is not essential to health".[21] Along with the adenoids, tonsils and spleen, the vermiform appendix is classified as part of the mucosal immune system. This is a secondary set of immune organs along the mucus-lined membranes of the body. This system is also called the mucosal-associated lymphoid tissue or MALT.[22] Although we can live without the vermiform appendix, for some people it may be a good idea to hold on to it. Dr R.A. Wheeler and Dr P.S. Malone[23] argue that "the presence of a normal appendix is fortuitous, for instance patients with spinal injuries and tumors, and those with chronic idiopathic constipation". They maintain that the vermiform appendix can be useful in reconstructive surgery, and claim that the removal of the vermiform appendix without serious reason "should be abandoned to preserve the appendix for future use should the need arise".[24-25]

IS IT TRUE THAT KING TUT'S PENIS IS MISSING?

(Asked by Glen Richman of Newark, New Jersey, USA)

Here's a fun one. It was there, then gone, then back again. It was missing, but it is no longer (so to speak). King Tutankhamun ruled ancient Egypt from about 1336 to 1327 BC. The discovery of his tomb in November 1922 is 1 of the greatest archaeological finds ever. It was not until November 1925 that King Tut's mummified body was actually photographed and autopsied. At that time his genitalia were noted as being present. His penis had been wrapped separately in an erect position, probably to give him fertility in the afterlife. His scrotum was swaddled next to his body. In 1968 it was decided that X-ray techniques not available in 1925 justified a re-examination of King Tut's body. It was then discovered that his penis was missing. Had it been taken as a souvenir? Was it merely mislaid? Was it secretly stolen to be used by some King Tut cult? In January 2005 a new autopsy was carried out — and his missing penis turned up again (so to speak).[27]

WILL WE EVER DEVELOP AN ARTIFICIAL KIDNEY?

Artificial kidneys of sorts have existed since at least 1944. But they are outside the body and have to be connected to it. A dialysis machine is an artificial kidney in the sense that it performs the role of a real kidney. Dialysis can be performed through a hospital dialysis unit, usually twice a week, or through a home dialysis unit, usually once a day. Today, the ideal alternative to long-term dialysis is transplantation of a kidney. The operation is technically quite easy, but there are 2 problems. First, there is a shortage of donors matched for histo-compatibility molecules. Second, there is the problem of kidney graft rejection by the recipient's immune system because of histo-incompatibility.

According to Dr V. Bonomini and colleagues at the University of Bologna, a fully portable, fully "bionic", fully artificial kidney is still some years away. However, in a process similar to that used for regenerating bone, Dr David Humes of the University of Michigan and Dr Anthony Atala of Harvard University are using kidney cells to make "neo-organs" that possess the filtering capability of the kidney. In an attempt to solve design problems with an artificial kidney,

Humes and others at the University of Michigan are experimenting with adding a "bioreactor" unit to a dialysis unit to be installed in the body. The bioreactor consists of many hollow and porous tubes combined with a single layer of proximal tube cells. Proximal tube cells are coiled and lined with cells, carpeted with tiny hair-like structures (villi) and stuffed with cell food (mitochondria). Ironically, these proximal tube cells are derived from pigs. A dialysis bath fluid is then flushed through the lumen (the cavity of the tubes) where molecules and ions can be picked up by the portion of the cell (the apical surface) exposed to the lumen. Discharge of essential molecules, ions, and hormones at the basolateral surface of the cells places these materials back in the blood — which is just what the kidneys do. So far, all the testing has been performed on dogs. The results seem promising.[27]

WILL WE EVER DEVELOP AN ARTIFICIAL BLADDER?

Replacement bladder segments are available if the loss of a functioning bladder occurs. Dr D. Rohrmann and colleagues at the University Hospital in Aachen, Germany, have developed a silicone rubber whole bladder prosthesis. Animal tests were completed in 1996. The doctors write: "The positive outcome of these animal experiments suggest[s] this system would be useful for human bladder substitution."[27, 28]

WILL WE EVER DEVELOP AN ARTIFICIAL PANCREAS?

An artificial pancreas is on its way too. Dr J. Jaremko and Dr O. Rorstad of the University of Calgary write that developments such as better miniaturisation of insulin pumps "are significant steps towards a clinically applicable artificial pancreas". Developments should be completed over the next few years.[27, 29]

Between 5 and 7 per cent of men under age 64 and 10 to 20 per cent of men over age 64 have urinary incontinence.

The average person ingests about a tonne of food and drink each year.[15]

～ CHAPTER 12 ～

OTHERWISE INSIDE

H.G. Wells once wrote, "I am a temporary enclosure for a temporary purpose; that served, my skull and teeth, my idiosyncrasy and desire, will disperse, I believe, like the timbers of a booth after a fair."

WHAT WEIGHS MORE: MUSCLE OR FAT?

(Asked by Laurie Perry of Matraville, New South Wales)

Muscle weighs more than fat. This is because the body cells that compose muscle are denser than the body cells that compose fat. Fat-cell cytoplasm (the substance that fills most of the cell) contains mostly oils (lipids), whereas muscle-cell cytoplasm contains mostly contractile proteins (myosin and actin) suspended in a watery environment. In general, lipids are less dense than water. You can demonstrate this by mixing salad oil with water in a bottle and shaking it. Eventually the oil will separate from the water and float to the top. This lower density of lipids makes fatty tissue less heavy than muscle tissue. It's really that simple.

HOW LONG DOES IT TAKE TO LOSE 1 POUND OF BODY FAT?

The answer to this depends upon you, your level of health, level of fitness, the type and amount of exercise you do, and so on. There are 3500 calories in a pound (450g) of body fat. If you calculate the number of calories you take in during a 24-hour period and the amount of exercise you do in that same 24-hour period to burn the calories, you can work out the ratio of calorie gain or loss. You can then figure out how long it will take you to lose 1 pound of body fat.

WHAT IS THE ENDOCRINE SYSTEM AND WHAT DOES IT DO?

(Asked by Sophie Brown of Lindfield, New South Wales)

The endocrine system consists of organs that release secretions (hormones) directly into the bloodstream. These hormones affect the working of cells. "Endocrine" comes from 2 Greek words, *endon*, meaning "within", and *krinein*, meaning "to separate". If this hormone delivery process happened on the outside of the body, it would be called the exocrine system.

In the past, the list of endocrine organs was brief. They included the pituitary, adrenal, thyroid and parathyroid glands (organs specialising in secretion), plus the endocrine pancreas and gonads. But now we know that many other organs make hormones. For instance, the heart makes a hormone that helps regulate blood pressure. Kidneys make hormones that also help regulate blood pressure. Fat cells make a hormone that regulates appetite. The thymus makes a hormone that controls the function of lymphocytes (the white blood cells that fight infection and disease) and other immune cells. All of these hormones bind to receptor proteins located in or on the surface of target cells in the body. When activated by hormones, these receptor proteins can cause alterations in target cells or in how our DNA is expressed. This produces new cellular proteins vital to survival.

WHAT HAPPENS IF THE ENDOCRINE SYSTEM FAILS TO WORK WELL?

If the endocrine system fails to work many conditions can result. For instance, dwarfism results from the insufficient production of growth hormone by the pituitary gland. Some of the symptoms are slow growth, short stature, abnormal body proportions, delayed sexual development, headaches, excessive urination and thirst.

WHAT CAN UPSET THE ENDOCRINE SYSTEM?

Almost anything can upset the endocrine system, including genetic problems, serious injuries, extreme physical or psychological stress, starvation, extreme exercise, infectious diseases, taking hormones, drugs or abusive substances, or exposure to harmful chemicals. The endocrine system is absolutely vital to the whole make-up of body activity, energy and change (metabolism).

WHAT IS THE PITUITARY GLAND AND WHAT DOES IT DO?

(Asked by Sophie Brown of Lindfield, New South Wales)

The pituitary gland is sometimes called the "master" gland of the endocrine system. This is because it controls the functions of the other endocrine glands. The pituitary gland is no larger than a pea and is located at the base of the brain. It is attached by nerve fibres to the part of the brain that affects it, the hypothalamus.

The pituitary gland itself consists of 3 sections: the anterior lobe, the intermediate lobe and the posterior lobe.

Each lobe of the pituitary gland produces specific hormones that do specific things. The anterior lobe produces growth hormone (prolactin) to stimulate milk production after giving birth, adrenocorticotropic hormone (ACTH) to stimulate the adrenal glands, thyroid-stimulating hormone (TSH) to stimulate the thyroid gland, follicle-stimulating hormone (FSH) to stimulate the ovaries and testes, and luteinising hormone (LH), also to stimulate the ovaries or testes. The intermediate lobe produces melanocyte-stimulating hormone to control skin pigmentation. The posterior lobe produces antidiuretic hormone (ADH) to increase absorption of water into the blood by the kidneys and oxytocin to contract the uterus during childbirth and stimulate milk production. Many authorities believe that abnormalities in hormone secretion by the pituitary gland greatly affect ageing. For example, in menopause, around age 55, the hypothalamus and pituitary stop stimulating the ovaries in women. This leads to abnormalities in bone strength, muscle strength, blood pressure, memory functions and other aspects of ageing.

COULD I LIVE WITHOUT A PITUITARY GLAND?

(Asked by Sophie Brown of Lindfield, New South Wales)

This is not recommended. If the pituitary gland were to be removed in childhood, the consequences would include not growing normally and not maturing sexually.

WHY DOES ALCOHOL MAKE ME DRUNK?

When you drink, alcohol is quickly absorbed by your body through the stomach lining and directly enters the bloodstream. Alcohol dilates blood vessels, causing a warm feeling. However, drinking alcohol actually causes the body temperature to drop, so this warm feeling is illusory and temporary. Alcohol also depresses the normal functioning of the central nervous system (CNS). Small quantities of alcohol depress the inhibitory mechanisms of the brain, producing sensations of relaxation and loss of inhibition usually mistaken for stimulation. Larger quantities of alcohol sedate and impair muscular coordination, speech and mental function in general. With still more alcohol, the effect spreads from the brain's frontal portions to the cerebellum. This can depress vital centres of the CNS and cause loss of consciousness and even death.

CAN DRINKING OR EATING ANYTHING SOBER ME UP?

No. This is a dangerous urban myth. This falsehood has probably cost many people their lives, not just their driver's licences, as they continue to think that they can drink and drive. There is no known way to sober up a person who is drunk. Time and rest are the only remedies to remove alcohol from the blood. Alcohol is carried in the blood to the liver, where it is broken down into acetaldehyde and other by-products that leave the body through the urine and, to some extent, the lungs. The elimination rate is about 15 millilitres (half an ounce) of alcohol per hour. There is no way to speed up this process, but the absorption process can be slowed down by drinking alcohol while food is in the stomach.

CAN COFFEE SOBER ME UP?

No. Drinking coffee might be helpful to relieve some symptoms of a hangover, since the caffeine in coffee is a stimulant and alcohol is a depressant. Also, any liquid might help a hangover because dehydration is one of the usual hangover symptoms. Alcohol actually removes water from the body. This is why you urinate more when you drink alcohol; it's not because you're drinking more liquid.

Alcohol and caffeine are both diuretics. They both cause you to urinate more and feel thirstier. The diuretic property of caffeine may cancel out any thirst-quenching benefit of coffee. Coffee serves no purpose in lowering blood alcohol levels.

IS THERE A CURE FOR A HANGOVER?

Everyone has their own idea about this one. Everything has been tried from orange juice to raw eggs, from aspirin to eating a greasy breakfast, from a Bloody Mary (more alcohol) to a "hair of the dog" (also more alcohol). The topic is almost ignored by researchers, who instead concentrate on the effects of alcohol, the nature of alcohol dependence and the elimination of that dependence. It is not even agreed among scientists as to what constitutes a hangover. From time to time, if not a cure, then a possible hangover helper is suggested by research. In 1 of these, an extract taken from the skin of the prickly pear, *Opuntia ficus* (OFI), has had moderate success. Researchers headed by Dr J. Wiese concluded in a 2004 study that "The symptoms of the alcohol hangover [nausea, headache, dry mouth, body aches etc] are largely due to the activation of inflammation. An extract of the OFI plant has a moderate effect on reducing hangover symptoms, apparently by inhibiting the production of inflammatory mediators."[1, 2]

WHY DO I SHIVER WHEN I'M COLD?

Shivering is part of your body's internal temperature control system. Behind your eyes is the hypothalamus. It controls body temperature. Shivering is 1 way the hypothalamus makes heat as your body gets chilly. It also makes you perspire when you're hot. When you're cold, your body constricts your skin's cutaneous vessels while at the same time it opens the vessels that return blood to your heart. Blood is drawn away from the body's exterior layers of skin and towards the body's interior. This is why you feel chilled. It is most important for your survival that your interior remain at the right temperature. If your surface is a little cold at times, you can live through it. Also as a means to produce heat, your thyroid tells your muscles to begin contracting, causing you to shiver. You can shiver if you've got an infection and also if you're frightened. In either case, your brain perceives your body as cold and acts accordingly.

WHAT IS THE LOWEST BODY TEMPERATURE AT WHICH I CAN SURVIVE?

Normal body temperature is about 37° Celsius (98.6° Fahrenheit). Taken rectally, a body temperature of less than 28° Celsius (82° Fahrenheit) makes it impossible to live.[3]

WHAT IS THE HIGHEST BODY TEMPERATURE AT WHICH I CAN SURVIVE?

Life-threatening heat injuries are sustained to the body at 43.3° Celsius (110° Fahrenheit).[3]

WHERE DOES MY BODY TEMPERATURE ORIGINATE?

(Asked by Kylie Cramer of Dynnyrne, Tasmania)

It was once thought that an organ, such as the heart or the brain, heated the blood and was thus responsible for body temperature. However, research and clinical observations have established that the origin of the body's heat is all cellular activities of the body combined. The brain's role is that of a general, while each cell's role is that of a soldier. As every living cell does its job, chemical reactions are continually occurring. But these reactions are not 100 per cent efficient and a waste product is produced in the form of heat. Muscle contraction also produces heat by friction. Shivering when you are cold is a good illustration of this. When the heat from metabolism is released, it is picked up by interstitial fluid, which enters the blood and is distributed to all parts of the body. The body has heat-sensing receptors that detect when the body is too hot or too cold. It then makes whatever adjustments are necessary to maintain the 37° Celsius (98.6° Farenheit) that is the normal internal temperature of the body.[4]

WHY DOES FAT UNDER THE SKIN KEEP US WARM?

Keeping fat deposits is the way humans and other animals store tissue that can be converted into energy if food is scarce. Unlike bone and muscle tissue, fat does not have another routine job. Thus, fat tissue deposits do not have very many blood vessels within them. Anything that slows down the heat escaping from the body helps keep the body warm. This is why wearing clothes keeps you warmer when it is cold outside. Clothes slow down the heat loss from your body. Fat layers perform the same work because of the low blood flow through them. The fat keeps some of your blood from getting too close to the skin where it can quickly lose heat. The result is almost like carrying a blanket around with you under your skin. There is a large, thick layer of fat just under the layers of muscle across your abdomen. There are a lot of blood vessels in your abdomen associated with your intestines and other internal organs. This fat layer is called the greater omentum. It is a very good use of body fat to preserve body heat.

Humans whose ancestors lived in cold climates have more body fat and the fat is distributed uniformly over much of their bodies. This better protects them from heat loss. By contrast, humans whose ancestors developed in hot climates have more localised fat deposits. This leaves large areas of skin with many blood vessels near the skin, which allows easier cooling off if they become overheated. Fat is the most concentrated source of energy of the 3 basic food types. Fat has 9 calories per gram; protein has 4 calories per gram; carbohydrates have 4 calories per gram. The human body burns some of its stored fat when it has not received enough food to supply all of its needs. Thus, not only does a fat layer help keep us warm by preventing heat loss to the air, but it is also 1 of the primary fuels that the body burns to produce the heat in the first place.

The only parts of the body where the skin has virtually no subcutaneous fat are the eyelids and the scrotum.[3]

WHY DO MEN SOMETIMES DEVELOP LARGE BREASTS?

Excessive enlargement of the male breast is called gynecomastia. The term is from the Greek *gyne* meaning "woman" and *mastos* meaning "breast". The breasts of some men can become just as large as they are in the average woman. The mammary glands within a man's breast can be fully functioning too. This means that a man could be capable of giving milk. Gynecomastia can be caused by obesity, the use of oestrogens or steroids, the presence of a tumour, or even as a symptom of disease. In an abnormal genetic condition called Klinefelter's syndrome, about half of sufferers have gynecomastia. They also are usually very tall, with very long legs and small testes, and are infertile. According to Dr Allen Rosen,[5] gynecomastia can occur at any age but is most common in ageing men. As a man grows into middle age, often his male hormone (testosterone) production reduces somewhat, while his female hormone (oestrogen) production remains the same. The result is that female sex characteristics such as large breasts can appear.

Men troubled by breast enlargement should see their physician. Corrective treatments for gynecomastia include liposuction or surgical removal of tissue. Often the cause of gynecomastia is unknown or something that the person has done to themselves. Temporary enlargement of the breasts is not unusual or abnormal in adolescent boys, even in pre-adolescent boys. Dr R. Einav-Bachar[6] and 3 colleagues reported that in their 2004 study, 5 per cent of 581 boys referred to them for the appearance of gynecomastia were pre-pubertal. In most cases gynecomastia was the only symptom and the cause was unknown or self-originating.[7,8]

WHAT MAKES BREASTS GROW?

(Asked by Amy Francis of Pendle Hill, New South Wales)

A female hormone called oestradiol makes breasts grow to mature size at puberty. It also makes them grow during pregnancy, especially during the first 2 trimesters. According to Dr Frederick Sweet,[9] oestradiol is made in the female ovaries and is transported by the bloodstream all over the body. However, only a few specialised tissues are stimulated to grow by the oestradiol made in the ovaries. In puberty, it is the development of the ovaries that triggers the growth of breasts. The cells in the breast that are responsive to oestradiol are those which are primitive or undifferentiated. However, after breast cells are differentiated into milk-producing cells, under the influence of oestradiol and other, still largely unknown, factors, normal breast cells can no longer be stimulated to multiply. The exception to this is breast cancer cells that can multiply throughout a woman's life.[10]

WHAT ARE THE LITTLE BUMPS AROUND A WOMAN'S BREAST CALLED AND WHAT DO THEY DO?

(Asked by Amy Francis of Pendle Hill, New South Wales)

The small bumps around the areola of the breast are called the tubercles of Montgomery. While the skin covering the breast is smoother, thinner, and more translucent than on most of the rest of the body, the areola's skin is even thinner and has complex sweat glands, hair follicles and sebaceous glands. The sebaceous glands secrete sebum. The tubercles of Montgomery are sweat and sebaceous glands that become more prominent in the second half of a woman's menstrual cycle. They also expand somewhat throughout pregnancy.

WHY ARE WOMEN'S BREASTS THE SHAPE THEY ARE?

Believe it or not, the shape of women's breasts is due to the size of the human brain. The large human brain has forced the human face to become flatter. The up-tilt, roundness and, to some degree, fullness of women's breasts evolved to prevent babies from being smothered when breast-feeding. Non-human primate babies and other mammals, with their protruding jaws, can suckle their flat-chested and small-breasted mothers without suffocating. But human babies, whose faces are flatter due to their large brains, both suck and breathe more comfortably from an up-tilted, rounder and fuller breast.[11]

WHY DO HUMAN FEMALES HAVE LARGE BREASTS WHILE FEMALES OF OTHER MAMMALS HAVE SMALLER BREASTS?

(Asked by Amy Francis of Pendle Hill, New South Wales)

It's always interesting to investigate why humans differ from other animal species. But as far as this question goes, science really doesn't know the answer. However, we do know that human females do not need large breasts to successfully breast-feed a baby (see Chapter 1). Adipose tissue in the breast is not required in large supply for normal lactation. Indeed, women nursing their babies for 6 months or more can undergo a reduction of breast volume to the pre-pregnant state and still produce an adequate amount of milk. The excess adipose tissue is undoubtedly related to the sex difference in overall adipose tissue that normally exists between men and women. Adipose tissue responds to oestrogen by increasing in volume.

WHY DO BREASTS SAG LATER IN LIFE?

This is one instance where women with small breasts have 1 up (or perhaps 2 up) on women with large breasts. They don't have to worry about sagging breasts because there's nothing much to sag. Breast ptosis is the medical term for sagging breasts. In fact, ptosis can refer to the sagging of any body part for any reason, including paralysis. For example, eyelid ptosis causes a person to be unable to blink or keep their eyelid fully raised; injury or stroke can result in eyelid ptosis.

Basically only muscle and skin hold breasts up. Breast ptosis is caused by a combination of at least 4 factors. The first is simply gravity, which pushes down on the breasts when you are standing or sitting, and, to a lesser extent, even when you are lying down. This has a drooping effect over time. The second is the ageing of breast muscle and skin as we get older; the muscle is weaker, the skin is thinner, so neither can "lift" as well as before. The third is the expansion and contraction of the breasts due to pregnancy and breastfeeding. The fourth is the flattening of the breasts due to the atrophying of breast glands after hormone levels have been altered due to menopause.

DOES WEARING A SUPPORT BRA AT NIGHT PREVENT BREASTS FROM SAGGING?

According to noted plastic surgeon Dr Michael Bermant, of Chester, Virginia, "Bras have little help for a sagging breast when sleeping. They can be a significant comfort for the massive breast [gigantomastia]. Stabilising excessive mass can lessen pull, tension and discomfort during normal sleeping movement ... Unless you sleep standing up, a support bra should not really help long term to prevent further sagging at night. Taking care of yourself, using a support bra during the day, and vigorous activities limit this problem to a certain degree ... Most breast ptosis is from loose skin suspensory igamentous tissue. The only non-surgical methods I have read about were in science fiction novels. External support ... may help lessen the problem getting worse. Nothing that I know of improves the situation without surgery."[12]

WILL WE EVER DEVELOP ENTIRE ARTIFICIAL BREASTS?

(Asked by Amy Francis of Pendle Hill, New South Wales)

Of course we've had breast implants for several decades now. Silicon-based, saline-based, soy-oil-based and other types of implants have been used to augment or fully reconstruct breasts. Dr Walter Holder and Dr Craig Halberstadt of the Carolinas Medical Center in Charlotte, North Carolina, are experimenting with techniques to grow soft tissue from a woman's own breast cells. This would be an alternative to current breast implants.[13, 14]

Hippocrates, the ancient Greek "Father of Medicine", believed that a flat-chested woman could enlarge her bust by singing loudly and often.

❧ ❧

Researchers claim that nobody really knows how frequently the average person takes a bath or shower. It also probably varies from season to season.[15–17]

❦ ❧

An average human being performing light work in a temperate climate loses nearly 2.4 litres (5 pints) of water a day. This water must be replaced or the person will die.[15-17]

❦ ❧

Even when no drink is consumed, ⅓ of a litre (⅔ of a pint) of water is taken in at each meal. About half of all ordinary food is water.[15-17]

❦ ❧

A man at rest is losing over 15 millilitres (half an ounce) of water through his skin every hour.[15-17]

❦ ❧

A woman has less water within her than a man of equal weight.[15-17]

❦ ❧

The maximum possible daily loss and replacement of water is about 24 litres (50 pints).[15-17]

❦ ❧

A 70-kilogram (156-pound) man holds 33 to 37 litres (70 to 80 pints) of water within his frame. About half of that will have been lost and replaced every 10 days.[15-17]

❦ ❧

Even under ideal conditions, a man without food or water will die when he loses 15 per cent of his body weight — usually within 10 days. With water alone he can survive for perhaps 2 months.[15-17]

∿∿ ∿∿

With water but no food, a 76-kilogram (168-pound) man can drop to less than 38 kilograms (84 pounds) and still live. Without any water, he'll be dead well before he has reached even 63 kilograms (140 pounds).[15-17]

WHAT IS THE LYMPHATIC SYSTEM AND WHAT DOES IT DO?

(Asked by Ken Scanlon of Emu Plains, New South Wales)

The lymphatic system and the cardiovascular system are closely related structures that are joined by a capillary system. This is important to the body's defence mechanisms. It filters out organisms that cause disease, produces certain white blood cells and generates antibodies. It is also important for the distribution of fluids and nutrients in the body, because it drains excess fluids and proteins so that tissues do not swell up. Lymph fluid is a milky body fluid that contains a type of white blood cell called lymphocytes, along with proteins and fats. Lymph seeps outside the blood vessels into the spaces of body tissues and is stored in the lymphatic system to flow back into the bloodstream. Through the flow of blood into and out of arteries, into the veins, through the lymph nodes and into the lymph fluid, the body is able to eliminate the products of cellular breakdown and bacterial invasion.

Two very large areas are of significance in this system. The first is the right lymphatic duct, which drains lymph fluid from the upper right quarter of the body above the diaphragm and down the midline. The second is the thoracic duct, which is a structure roughly 40 centimetres (16 inches) long located in the mediastinum, and helps drain the body of unnecessary fluids. (The mediastinum is found in the chest's thoracic cavity, which also contains the heart, great vessels of the heart, oesophagus, trachea, thymus and the lymph nodes of the central chest.) It is through the actions of this system, which includes the spleen, thymus, lymph nodes and lymph ducts, that our body is able to fight infection and ward off foreign invaders. Lymph fluid plays an important role in the immune system and in absorbing fats from the intestines. The lymphatic vessels are present wherever there are blood vessels and transport excess fluid to the end vessels without the assistance of any "pumping" action. There are more than 100 tiny oval structures called lymph nodes throughout the body. These are mainly in the neck, groin and armpits, but are scattered all along the lymph vessels. They act as barriers to infection by filtering out and destroying toxins and germs. The largest body of lymphoid tissue in the human body is the spleen. According to Dr Brett Ellis,[18] the lymph nodes are filled with immune system cells that help fight off infection. As such, the lymph nodes can help prevent cancer. Unfortunately, they can also be responsible for spreading cancerous cells as the lymph fluid is returned back to collecting ducts.

WHAT IS A MUSCLE CRAMP?

(Asked by Jim Crossman of Carlton, Victoria)

A cramp is an exaggerated muscle contraction that is often quite painful. There are many ways in which cramps can be triggered. A cramp can result if something goes wrong with a nerve, the muscle's electrical activity, or the biochemical interactions within the cells of the muscle. Physiologists still don't know what causes cramps in all cases, but there are at least 5 theories: (1) cramps can occur due to lactic acid build-up after strenuous exercise; (2) cramps can occur due to dehydration after overheating and strenuous exercise; (3) cramps can occur when not enough oxygen gets to the muscle — this is related to lactic acid build-up, since such a build-up happens during anaerobic activity; (4) cramps can occur from an electrolyte imbalance involving calcium, potassium or sodium ions; and (5) cramps can occur when a muscle in a "shortened" position causes the spinal control mechanisms to overload the muscles by turning off what is termed the Golgi reflex activity.

At the risk of being too technical, normal muscle contraction is the result of a series of complicated biophysical and biochemical steps. An electrical message is transmitted to and from the brain, up and down a nerve, and to and from the cramp site in a muscle. The nerve releases a neurotransmitter onto receptors in the muscle. This causes a change in the distribution of charged ions on either side of the muscle membrane, which triggers proteins in the membrane to release calcium into the muscle. This in turn causes biochemical interactions between filamentous proteins involved in muscle contraction and the energy those proteins need to function normally. But whatever the cause, cramps are painful!

WHY DO MUSCLES EXPAND AND CONTRACT?

(Asked by Charles Bernie of Lapstone, New South Wales)

Most people believe that muscles expand and contract (or push and pull) in order to move various parts of the body. But this isn't true. Muscle tissue has the uniform property of only being able to contract, or shorten, when stimulated by the nervous system. When the brain sends impulses to a leg muscle, for instance, that leg muscle shortens.

Muscles or muscle groups that control joints are designed to work in opposing pairs. For example, when the quadriceps muscle group at the front of the thigh contracts, the leg below the knee extends. Bending the knee is then a matter of contracting the opposing muscles that run down hamstrings at the back of the thighs. When one muscle group contracts, the opposing muscles relax so that they can be pulled back into their elongated form.

Muscles of the arms, legs and any other part of the skeletal system that is capable of movement are referred to as voluntary (striated) muscles. They are the only type of muscle in the body that is under voluntary control. Another way of explaining this is that these skeletal muscles are attached at both ends to bones, cartilage, ligaments, skin or other muscles. When a muscle contracts to move part of the body, the attachments on the other end remain static. But when moving that part back in the opposite direction, this procedure is reversed, with a different set of muscles contracting while the first pulling muscles remain inactive. At no time does any muscle that has just pulled ever push back in the opposite direction.

HOW TALL WOULD YOU BE IF YOU CONNECTED ALL THE MUSCLES OF THE BODY?

(Asked by Ian Bolger of Battery Point, Tasmania)

This is a very difficult question to answer. There are many types of muscle in the body. For example, the heart is a muscle. How long is the heart? Some muscles are very small, almost microscopic, like the stapedius muscle which guards one of the small ear bones. Other muscles — sphincters — are round and guard the size of an opening. For example, a sphincter muscle controls the opening of the iris of the eye. How would you include these types of muscles in your body, length estimate? It's important to think of a muscle as not really being a separate organ. Instead, muscles are made up of fibres that can act alone or together. In order to determine the length of a muscle, you would have to separate out the fibres and lay them end-to-end, since these muscle fibres are separate structures with all the actions of a muscle. If you did that, you might find that the line stretched on for many kilometres.

WHAT IS THE LONGEST NERVE IN THE BODY?

(Asked by Sandra Carlson of Red Hill, Queensland)

The best candidate for this distinction is the sciatic nerve, which begins at about the middle of the buttocks and extends down the back of the thigh, deep into the hamstring muscles. Just above the knee, it splits into the tibial nerve and the common fibular nerve. According to Dr Stephen Moorman,[19] the tibial nerve is longer than the common fibular nerve. This is because 1 of its branches, the medial plantar nerve, extends onto the sole of the foot and all the way to the end of the big toe.

If the sciatic nerve is the longest nerve, it's due to its axon branches. Axons are nerves that send sensory messages from the spinal cord all the way to the big toe. The axons in the sciatic nerve are the longest in the body. The sciatic nerve innervates all the muscles of the posterior thigh and all the muscles of the calf and foot through its 2 branches, the tibial and fibular nerves. The tibial and fibular nerves also carry sensory information from most of the skin of the calf and foot.

It's a dead heat between the 2 other candidates for longest nerve, the median nerve and the ulnar nerve. These nerves both extend from the shoulder to the wrist. The distance from the shoulder to the wrist is longer than from the hip to the knee (the length of the sciatic nerve). It is also greater than the distance from the knee to the ankle (the length of the tibial nerve). All the muscles of the front of the forearm and all the muscles of the hand are innervated by either the median nerve or the ulnar nerve. These 2 nerves carry sensory information from some of the skin of the forearm and most of the skin of the hand.

So what's the longest nerve? Some experts judge 1 way, others another. They can't agree. The nerve!

CAN I LOSE WEIGHT BY THINKING A LOT?

(Asked by M. Marsh of Napa, California, USA)

Thinking, no matter how much and how hard, is not a very efficient way to lose weight.

The brain constitutes about 2 to 3 per cent of the body's mass, but it consumes about 20 per cent of the body's energy intake. The brain is very energy-hungry indeed. Yet thinking burns up very few calories. It has been estimated that thinking for an hour burns about ⅕ of a gram of fat. This is not a great amount compared with, say, riding a bicycle for an hour. The act of thinking is a relatively small part of the brain's many important functions. Most of the calories burned by the brain support these non-thinking activities. The brain plays a vital role in regulating the internal systems of the body (homeostasis); cardiac and respiratory control; body temperature regulation; body posture control; sensory input from numerous systems, including the eye, the ear, the nose and the tongue; stretch receptors in the muscles; position sensors (proprioceptors) in the joints; and pressure and pain receptors in the skin. The arterial system control mechanism of the brain, which is crucial for the maintenance of blood pressure and circulation, needs to send motor signals to innumerable blood vessels. The brain must also regulate its own chemistry. For example, the right neurotransmitters and enzymes must be released at the right time and in the right quantity. Much of this brain activity occurs below the level of consciousness. Doing all of this requires vast sensory input to the brain, complex processing and an equally large sensory output to produce the desired outcome. To do all of this, the brain possesses 10 trillion nerve cells. It's great that the brain can do all of this for so few calories.

WILL WE EVER DEVELOP AN ARTIFICIAL LIVER?

Whole organ transplantation is usually regarded as the only clinically effective method of treating chronic liver failure. But donor shortage and the usual problems of transplantation (rejection, lack of donors etc) have stimulated the development of an artificial liver. Several versions are currently being clinically tested around the world. One is being developed by Dr Y. Kamohara and colleagues at the Cedars-Sinai Medical Center in Los Angeles. The Kamohara team reports that their artificial liver is showing real promise. But they add that such an innovation is only possible due to, among other things, the recent "availability of new biomaterials".[20]

Doctors at the University of Michigan are also testing an artificial liver of sorts. It temporarily takes over at least some of the functions of a failing liver. According to Dr Robert H. Bartlett of the University's C.S. Mott Children's Hospital, tests are underway of an artificial liver system that is designed to work for patients who are waiting for a transplant. The HepaLife Company of Vancouver, Canada, is currently in the process of developing this first-of-its-kind artificial liver device designed for use by patients with liver failure.

However, what is called split liver transplantation has been successfully experimented with at Stanford University Medical Center in California, Mount Sinai Medical Center in New York and elsewhere. In split liver transplantation, a healthy liver is divided up among compatible patients, including children. It is sometimes called a "2-for-1 transplant". One of the first reports of success with this procedure was in 1996.[21]

So there are a lot of things happening. It is just a question of when the research, product development and testing will be completed.[22, 23]

CHAPTER 13

SLEEP

The great Miguel de Cervantes once wrote: "Blessings on him that first invented sleep! It covers a man, thoughts and all, like a cloak; it is meat for the hungry, drink for the thirsty, heat for the cold, and cold for the hot. It is the current coin that purchases cheaply all the pleasures of the world, and the balance that sets even king and shepherd, fool and sage."

WHY DO WE SLEEP?

Sleep. We spend ⅓ of our life doing it, but science is puzzled about why. You might think the answer is obvious: we sleep to recharge our batteries. But research suggests it's not that simple. The medical literature reveals dozens of cases of people who hardly ever sleep, yet are perfectly healthy. Indeed, often they're healthier than most of those who sleep the normal 8 hours per night.

A few years ago, British researchers reported a case of a 70-year-old woman who slept an average of 1 hour per day without any naps. She once stayed awake for 56 hours straight — and then only slept for an hour and a half. She felt fine and functioned normally. Then of course there's the most famous self-imposed insomniac of all time: Leonardo da Vinci. He slept only 15 minutes every 4 hours around the clock. So why do the rest of us spend 8 hours a day sleeping? And if we don't get our proper dose of sleep, why are we so cranky and out-of-sorts? Here are a few theories:

☞ Sleep keeps us out of danger — early humans needed protection from predators, most of whom hunted at night. Sleep reduced the boredom of hiding in caves.

☞ Sleep helps us save energy — sleeping part of the day allowed early humans to get by on less food. When we sleep, our body temperature drops a few degrees and we burn up less energy. This means that we don't have to eat as much. In prehistoric times, scarcity of food and starvation were always a possibility.

☞ Sleep gives our body time to recover from the stress of everyday life — by the time night arrives, our brain needs a break. Some scientists believe that during sleep our body is repaired. The brain repairs its cells and makes more of the chemicals it needs for brain activity during consciousness.

☞ Sleep helps memory — gives our brain time to sort out the sensory input of conscious activity and classify this information in our memory.

☞ Sleep helps us forget — some scientists claim that sleep helps our brain "unlearn" things. This is essential so that the brain doesn't become overcrowded with unnecessary information.

☞ Sleep allows us to dream — we dream only during REM sleep (rapid-eye-movement sleep). If we didn't sleep, we couldn't dream. Although most of us assume the reverse, some scientists argue that our conscious hours are really only preparation for our unconscious hours. After all, what we do in our dreams is often far more fascinating than what we do when awake. If total freedom is the ultimate human desire, and if dreaming is when we can be the most free, then dreaming is the most desirable human state.[1]

CAN I CAUSE ANOTHER PERSON TO DREAM?

Studies show that you *can* bring about a dream in another person. One way is by holding an open bottle of perfume under the sleeper's nose. Another is by whistling. A third way is by blowing air across the sleeper's face with a fan. You can also affect the content of a sleeper's dream. For example, turning on a light produces happier dreams, and darkening a bright room can induce nightmares.[2]

WHAT WOULD HAPPEN IF I WERE PREVENTED FROM DREAMING?

In a study conducted by Dr William Dement, a Professor of Psychiatry at Stanford University, it was found that dream deprivation produces anxiety, irritability, an inability to concentrate, paranoia, a ravenous appetite, depression and even suicidal thoughts. The study was very simple. Volunteers were assigned beds in a laboratory. Researchers observed them closely for 5 nights. Each time a subject's rapid eye movements indicated they were starting to dream (REM sleep), the researchers would wake them up and keep them from dreaming. It seems that it only takes a few days of dream deprivation before the symptoms begin to occur.[2]

WHY CAN'T I REMEMBER ALL OF A DREAM?

How much of a dream we remember depends upon how soon we awaken after the dream. In 1 study, sleeping subjects were whoken up while they were in REM sleep. Other sleeping subjects were woken up 15 minutes after they were in REM sleep. In 152 out of 191 cases, the subjects awakened during REM sleep remembered they were dreaming and could recall a great deal of the content of their dreams. Of the subjects awakened 15 minutes after their REM sleep, hardly any could remember dreaming at all. If you're woken during or shortly after REM sleep, you're probably more likely to remember more of your dream.[2]

WHAT MAKES A DREAM?

Psychological research and clinical experience reveal that dreams are produced by a complex interaction of your mind, your emotions and the environment in which you sleep.

WHEN DID I FIRST DREAM?

There is evidence that the foetus dreams. Researchers place this time at approximately 23 weeks after conception, or about 15 weeks before we are born. Indeed, REM sleep is virtually the only type of sleep the foetus engages in. It is only at 36 weeks that non-REM sleep is detected. This is about 2 or 3 weeks before birth. Therefore, you could say that between the ages of 23 and 36 weeks, whenever the foetus is sleeping, it's dreaming.

DO WE SLEEP AND DREAM MORE OR LESS AS WE AGE?

Researchers have found that the younger we are, the larger percentage of our sleep time is dream sleep. Infants spend more time in dream sleep than children, children more than young adults and young adults more than the elderly. About 4 million Australians each year suffer from insomnia, which can lead to serious sleep deficits and health problems. Insomnia tends to increase with age and affects about 40 per cent of women and 30 per cent of men. The medical literature reveals that some chronic insomnia patients have gone without sleep for 5 years and even longer.

HOW IMPORTANT ARE OUR DREAMS TO THE REST OF OUR LIVES?

Prior to Freud, we didn't take dreams too seriously. Freud discovered all sorts of importance in our dreams and we've never looked back. Now, researchers are investigating all sorts of aspects of the dream–wake state. One of these is hypnagogia. Hypnagogia (from the Greek word *hypnos* meaning "sleep" and *agogos* meaning "leading" or "leader") is the strange semiconscious state we are in just before falling asleep and just before waking up. According to Dr Andreas Mavromatis, in the classic work on this topic,[3] by studying this unique semiconscious state we can understand many aspects of our conscious and unconscious lives. These include dreams, hypnosis, hallucinations, psychic phenomena, near-death experience, creativity, schizophrenia, the overcoming of pain, meditation and self-healing. Also, those uncontrollable and startling body jerks that we experience just when we are falling asleep or waking up are called hypnagogic jerks. These illustrate what a strange state the body is in when it comes in or out of sleep.

HOW LONG DOES A DREAM LAST?

(Asked by Nicole Murdoch of Gordon Park, Queensland)

According to Dr Max Hirschkowitz from Baylor College of Medicine in Houston, "dreams actually occur in real time".[4] A typical dream lasts 10 to 40 minutes, though it was once thought that dreams occurred in a flash. This idea probably began with an observation by a lawyer during the time of the French Revolution. He awakened from a dream of being beheaded to find that the headboard of his bed had fallen on him. The lawyer deduced, incorrectly as it turned out, that the entire prolonged dream happened in the brief moment that he was hit and awoke. We know now that we dream about things that are on our mind — fears, worries, anticipated events and so on. We also know that to remember a dream, you usually have to wake up. Thus, the lawyer was not only a good sleeper, but also should have been worried about losing his head. The guillotine was, of course, a popular and infamous method of execution in revolutionary France. Adds Dr Hirschkowitz, "I would speculate that he had that dream most every night but this time he was awakened from it and therefore remembered it."[5]

WHAT PART OF THE BRAIN IS RESPONSIBLE FOR DREAMS?

Dreams are probably the most interesting things we humans do during sleep. Research undertaken by Dr Thomas Balkin and Dr Allen Braun[6] has attempted to pinpoint the precise location/s in the brain where the greatest activity takes place during REM sleep, when dreams occur. They used positron emission tomography (PET) scan data to measure blood flow in the brain. When the brain is in a normal waking state, the prefrontal cortex just behind the forehead is very active. This is the most highly evolved part of the brain, the seat of intelligence, reasoning, short-term memory and self- reflection. During REM sleep the prefrontal cortex completely shuts down. Instead, what is most active is the limbic system, the part of the brain that controls emotions, senses and long-term memory. Dr Balkin and Dr Braun say, "This might account for the heightened emotional tone during REM dreaming, the sexual, aggressive, anxious effects, as well as the retrieval of long-term memories."[7]

Some have speculated that the fact that short-term memory is deactivated may account for the bizarre content of dreams. It also might account for the amnesia typically connected with dreams, i.e., why we do not remember them very well. Unlike conscious thoughts, dreams are not properly encoded. Working memory is not active. Balkin and Braun's studies have also shown that during sleep the primary visual cortex (what we use to see when we are awake) is inactive. Instead, what is operating is the extrastriate area of the brain. This is the "visual" area of the brain, which processes complex objects like faces. It also is involved in emotions. The role of the extrastriate area may help to explain the often "florid visual imagery" of dreams. Dr Mark Solms, a neurologist at St Bartholomew's Hospital in London, contends that "We don't really know what the functions of dreams are. We don't know the mechanism that makes us dream like we do. But my research shows psychological mechanisms are essentially involved in the construction and form of dreams. I think we're on the threshold of a very exciting era of dream science."[8]

WHAT IS SLEEPING SICKNESS AND WHY CAN'T PEOPLE WAKE UP?

Sleeping sickness is caused by a protozoa parasite transmitted by the bite of the tsetse fly. The parasite multiplies slowly in the body, causing fever and malaise that eventually end in coma and death. Sleeping sickness is almost always fatal if left untreated. As such, it remains a threat to rural communities, but is seldom a problem for urban dwellers.

The fact that a sufferer cannot be awakened from their deep coma is why the disease got the name "sleeping sickness". The first symptoms are difficult to recognise. Often all that is apparent is a small skin lesion (chancre) at the site of the bite, which erupts within a few days. The full-blown disease can last 2 to 3° months. As the diseases progresses, the sufferer can experience fever, extreme fatigue, headache, lack of appetite, skin rash, joint swelling, conjunctivitis (inflammation of the clear membrane that coats the inner eyelid and outer surface of the eye), myocarditis (inflammation of the heart walls), heart aneurysms, congestive heart failure, coma and eventually death. Infants can develop meningitis as well. The medical term for sleeping sickness is trypanosomiasis. Human African trypanosomiasis (HAT) plagues sub-Saharan Africa. HAT comes in two forms depending upon which parasite infects the sufferer: *Trypanosoma brucei gambiense* affects western and central Africa while *Trypanosoma brucei rhodesiense* affects eastern and southern Africa. In all, more than 66 million people in 36 countries of sub-Saharan Africa are exposed to HAT. In its website warning to travellers, the Centers for Disease Control and Prevention in Atlanta report that the World Health Organization recorded 45,000 cases of HAT in 1999, "however the actual prevalence of cases is estimated to be between 300 000 to 500 000". The CDC add that there is no vaccine against HAT, but treatment is available and usually successful if the disease is caught early.

WHY DON'T HUMANS HIBERNATE?

Hibernation is a state of inactivity in an animal brought about by short days, cold temperatures and limitations in food. Some animals survive winter when they migrate over distance. Other animals "migrate over time" by hibernating. Hibernation is a survival strategy to conserve energy. True hibernation (also known as deep hibernation) is when an animal is inactive for weeks or months. Their body temperature can drop to 5° Celsius (41° Fahrenheit). Torpor (also known as shallow hibernation) is when an animal is inactive for a much shorter period of time, say, only during the cold hours of each night during winter. Their body temperature drops to no lower than 15° Celsius (59° Fahrenheit). The big disadvantage of hibernation is that the animal is defenseless when they hibernate. Thus, a safe and secure hibernating place (hibernaculum) is essential.

There are at least 2 reasons why humans are not good hibernators. First, humans do not always live where a hibernaculum is available, e.g., in a desert environment. Second, animals that hibernate well have a high amount of brown fat in their bodies. Humans do not. Brown fat is a type of adipose (fat) tissue that is very good at conserving energy. By contrast with white adipocytes (fat cells), brown adipocytes contain several small membrane compartments (vacuoles) for storing energy, more capillaries for transporting oxygen and far more of the cell's power plants (mitochondria) for making energy. Brown fat is present in human newborns. About 5 per cent of their fat is brown fat. It is located in the shoulders and back and along the spine. As we mature, we drop our brown fat. Some adults retain some brown fat, but not nearly enough to hibernate — although some try to, especially when working.

In February 2002, Dr Peter Morgan and colleagues at the Rowett Institute in Aberdeen, Scotland, discovered the gene responsible for hibernation.

❧ ❧

It was previously believed that no primate hibernates, including humans. But in June 2004, Dr Gerhard Heldmaier and colleagues from Phillips University in Enid, Oklahoma, discovered one primate that does: the fat-tailed dwarf lemur of Madagascar. Researchers are puzzled as to why this lemur, our primate relative on a distant branch of the evolutionary tree, needs to hibernate. Madagascar has mild winters, with a rich food supply for the lemur all year long.

WHY DOES MY FIANCÉ SLEEP WITH 1 EYE OPEN?

(Asked by Nicole Murdoch of Gordon Park, Queensland)

People often sleep without shutting their eyes. In sleep deprivation experiments, it was found that subjects could fall asleep although their eyes remained open. Parents report that they see this happen in their young children on occasion, especially when their children are overtired.

Some people normally sleep with 1 eye or both eyes slightly open. They seem to suffer no adverse effects from this. It usually has to do with the muscles of the eyelid, which can exhibit various degrees of relaxation. Of course, your fiancé could have a slight abnormality in the muscles of the lid or of the lid itself. You can always have him see a doctor, but if he sleeps normally and is healthy in every other way, there is probably no reason for this.

One further thought: perhaps your fiancé keeps 1 eye open because he wants to keep an eye on *you*?[9]

WHAT IS "SLEEPSEX"?

"Sleepsex" describes the condition in which a sleeping person commits sexual acts on themselves or someone else. It is related to sleepwalking (somnambulism). The term "sleepsex" was coined by Dr David Rosenfeld and Dr Antoine Elhajjar. In their article,[10] Rosenfeld and Elhajjar reported on 2 cases to illustrate their concept.

The first involved an individual who habitually engaged in sex *and* eating while sleeping. The second involved an individual who committed sexual battery while sleepwalking. Rosenfeld and Elhajjar conclude that

"Sexual behavior in sleep may be pleomorphic [having different forms during different periods of a person's life] and more common than realized in both the [psychiatric] patient and normal populations." The first book on sleepsex[11] was written by Dr Michael Mangan, a psychologist at the University of New Hampshire. The book presents an analysis of 60 fascinating and diverse cases. In 2002, Stanford University researchers, led by Dr Christian Guilleminault of the Sleep Disorders Center, analysed the cases of 11 patients in great detail and clearly illustrated the wide variety of sexual behaviours associated with sleepsex.[12] The Stanford team writes, "Complaints included masturbation, sexual assaults and continuous (and loud) sexual vocalisations during sleep." One patient attempted to strangle his wife while raping her as he slept. Another was charged with raping a woman while he was asleep, yet he was acquitted because he did not know what he was doing when he did it. The Stanford team adds that sleepsex behaviour brings about feelings of guilt, shame and depression in those who experience it. Another of their findings is that the sleepsex patients and their bed partners "often tolerated the abnormal behaviour for long periods of time without seeking medical attention". Perhaps the most fascinating finding of all was that sleepsex patients have "morning amnesia". They cannot remember the sexual acts they have committed the night before.[13]

WHAT ARE THE MOST COMMON SLEEP DISORDERS?

Scientists who study sleeping now recognise 78 sleep disorders that can rob us of our sleep and thus our health and vitality. According to the American Sleep Disorders Association and the American Academy of Sleep Medicine, here are a few of them:

☞ Idiopathic insomnia — this is a lifelong inability to get adequate sleep. It has no observable cause. Sleep experts assume that this difficulty is due to an abnormality of sleep–wake control systems in the brain. It may also be due to a problem in the sleep-inducing and sleep-maintaining systems or hyperactivity in the body's arousal system.

☞ Periodic limb movement disorder — PLM disorder occurs when the sleeper periodically moves a limb (usually a leg) in exactly the same way over the course of the night. A typical movement would be a kick or flex of the leg every 10 seconds. These movements disrupt sleep and lead to insomnia and daytime sleepiness.

☞ Post-traumatic hypersomnia — this is excessive sleepiness that develops as the result of physical injury or disease in the central nervous system. It can be caused by brain injury, neurosurgery, infection or spinal cord injury. The hypersomnia usually goes away over weeks or months.

☞ Restless leg syndrome — this syndrome is characterised by uncomfortable feelings (e.g., tingling, itching, crawling, pulling, or aching) in the leg or legs right before falling asleep. Such feelings are relieved by moving the leg/s, but return when the moving stops. This obviously interferes with falling asleep and can cause severe insomnia. Usually patients accumulate large sleep debts after many nights of restless legs until the resulting powerful sleepiness overcomes the unpleasant feelings and the patient sleeps. However, once enough sleep debt is worked off, the feelings once again interfere with sleep.

☞ Nocturnal leg cramps — as the name suggests, these are leg cramps, usually in the calf, that occur spontaneously during sleep. They may last for only a few seconds or as long as 30 minutes. The cramps cause arousal and disturb sleep. Their cause is not well understood.

☞ Confusional arousal — also called sleep drunkenness or sleep inertia, this disorder is an extreme example of the slowness most people feel upon waking. People with confusional arousal respond poorly to commands or questions and they have major memory impairment of things that have just happened or happened a short time before. Behaviours are often strange, such as picking up a lamp and talking into it because the person believes it is a phone. These typically occur when someone is aroused from a deep sleep in the first part of the night.

☞ Sleep starts — sometimes called hypnagogic jerks, these are sudden, brief contractions of muscles in the head, arms or legs that occur just as people are falling asleep. Sleep starts are experienced by most people at some time. When they are very strong or frequent they can lead to insomnia. Sleep starts can occur with such force and power that bed partners have been known to be knocked out when hit in the head by a jerking arm.

☞ Sleep talking — this problem can be precipitated by emotional stress, fever or other sleep disorders such as night terrors or sleep apnoea. Although it may bother bed partners, sleep talking is usually benign. The talk is usually brief and devoid of emotional content. However, sometimes it can consist of a long speech or be infused with anger and hostility. Sleep talking can be spontaneous, or induced by conversation with the sleeper.

☞ Sleep-related abnormal swallowing syndrome — people with this disorder have inadequate swallowing of their saliva while sleeping. Saliva builds up in the mouth, then flows down the throat and is breathed into the lungs. This causes choking and coughing and wakes up the sleeper.

☞ Sleep bruxism — is the grinding or clenching of teeth during sleep. The sound of grinding teeth can be unpleasant to bed partners who hear it. It can cause excessive tooth wear in the individual, too. It can also lead to jaw pain and headaches while awake.

☞ Impaired sleep-related penile erections — in men, erections are a natural part of REM sleep. When REM-related erections are not present, it indicates a physical cause of impotence. This can be a useful way of differentiating between psychological and physiological impotence.

☞ Sleep-related painful erections — sometimes the erections associated with REM sleep can be so intense as to be painful. This may cause night-time awakenings during REM sleep and subsequent sleep loss.

☞ Sleep paralysis — sleep paralysis is a common part of REM sleep, but is a disorder when it strikes outside of REM sleep. Usually, people with sleep paralysis are unable to perform voluntary movements either right before they go to sleep or upon waking in the morning. Sleep paralysis most often lasts for several minutes and then disappears.

☞ Sudden unexplained nocturnal death syndrome — as the name suggests, this syndrome is typified by the sudden death of healthy young adults while they sleep. Neither clinical history nor autopsy provides an explanation for death. The first signs are laboured breathing, gasping and choking. But the disorder is not sleep apnoea. Spasm (fibrillation) of the heart muscle has sometimes been detected. This syndrome has been most often seen in Southeast Asian men between the ages of 25 and 44.[14–16]

IS THERE ANY TRUTH TO THE BELIEF THAT SLEEPING NEXT TO A YOUNG WOMAN REJUVENATES AN OLD MAN?

(Asked by Glen Richman of Newark, New Jersey, USA)

This is a strange question. The practice is called Shunamitism, a widespread practice in earlier times. It was based upon the belief that the non-sexual sleeping with a young woman will rejuvenate an old man. It was also widely practised in cultures where there was a belief that one's life, spirit and health are contained in one's breath. The first recorded case of Shunamitism involved King David (1090–1015 BC) in the Old Testament. A young girl, Abishag of Shunem (from whom we get the name), was brought to lie with the elderly king in a non-sexual embrace.

According to Dr T.Vago, it was believed that the breath of someone very much younger is warmer and hence more life-giving than the breath of someone very much older. Shunamitism steadily fell from favour after the 17th and 18th centuries.[17, 18]

WHAT IS THE BEST WAY TO FALL ASLEEP AND STAY ASLEEP?

Forget counting sheep, chamomile tea and warm milk with butter; listening to music at bedtime is the secret to nodding off fast and getting a restful night's sleep. At least, that's what 2 Taiwanese researchers claim. According to them, the best kind of music is soft, slow tunes, such as light jazz, folk or orchestral pieces that have 60 to 80 beats a minute. The researchers even have a name for it: sedative music.

Believe it or not, the music will actually cause physical changes in your body. One of these is the lowering of the heart and respiratory rates to those conducive to a good night's sleep. The researchers studied the sleep patterns of 60 people between the ages of 60 and 83, all of whom had difficulty sleeping. Half — the music group — were given relaxing music to listen to for 45 minutes at bedtime, and half — the control group — were given no music and no other help to fall sleep. The results were that those who listened to a selection of soft, slow music experienced physical changes that aided restful sleep.

According to the head of the study, Dr Hui-Ling Lai,[19] "The difference between the music group and the control group was clinically significant. The music group reported a 26 per cent overall improvement [in sleeping] in the first week, and this figure continued to rise as they mastered the technique of relaxing to the sedative music. In addition, they suffered less dysfunction during the day since they had slept better the night before. Best of all, there were no side effects or pricey bills for prescription medicine."[20] Another study by the same research team found that preterm infants also slept better when they were exposed to sedative music.

In a US survey, it was found that 41 per cent of people sleep in a double bed, 31 per cent in a queen-sized bed, 21 per cent in a king-sized bed and 6 per cent in a single bed. But that only adds up to 99 per cent. And what about those who sleep in bunk beds?

∽ ∾

Insomnia strikes the elderly more than any other group. But there's good news. According to Dr M.

Woodward of the Austin and Repatriation Medical Centre in Melbourne, "With appropriate management, most older people with insomnia will experience improved sleep."[21]

Research shows that you remember what you learn just before going to sleep better than what you learn at other times.

Depriving someone of sleep has been used as a particularly cruel form of torture: people will eventually go mad. They first become cross, irritable, clumsy and forgetful, they then experience hallucinations.

Although some researchers dispute this, the less you sleep, the more you eat. You'll probably be at least a little hungrier on the day after you've been deprived of your usual full night's sleep.

Those suffering from the rare condition of chronic colestites experience total insomnia. This stops them from getting any sleep at all. The medical literature shows that some patients have gone without sleep for 5 years or more.

The human body can last longer without food than without sleep. You can go without food for several weeks, but after 10 days without sleep — that's total lack of sleep, without 1 catnap — you'll die.

Medical researchers have found that sleeping on your right side improves digestion.

❦ ❦

Pioneering studies at the University of Chicago found that dream periods average 20 minutes and can last up to an hour. They also found that the average adult spends about 22 per cent of their sleep time in dreams. Subsequent research has modified this view somewhat.

∽ CHAPTER 14 ∽
ENDINGS

The great Francis Bacon wrote: "There is no passion in the mind of man so weak, but it mates and masters the fear of death. And therefore death is no such terrible enemy, when a man hath so many attendants about him that can win the combat of him. Revenge triumphs over death; love slights it; honour aspireth to it; grief flieth to it." Pretty much sums it up, doesn't it?

THE MYTHS OF OLD AGE
Myth 1: Mental decline is a part of normal ageing
Researchers once believed that senility was inevitable, but this is no longer the case. According to Dr David Mitchell, a psychologist at Southern Methodist University in Texas, "the idea that memory inevitably deteriorates as you age came from studies that only tested one kind of memory. Now we see that there are multiple memory systems, and they each hold up differently as you age".[1] In fact, some types of memory improve with age. Research by Dr Mitchell found that subjects aged 57 to 83 scored higher on a vocabulary test than subjects aged 18 to 34. According to Dr Lydia Bronte, a psychologist at the Stokes Institute in New York, "most senility is really caused by Alzheimer's

disease and is unrelated to old age. Extraordinarily, only 2 of the people I interviewed complained that their memories weren't as good as they used to be. And, in fact, many memory problems are caused by vitamin deficiencies — particularly of the B complexes — and have little to do with aging." Dr Bronte adds that "aging leads to an increase in what psychologists now call crystallized intelligence — a capacity to apply knowledge to real life situations — what was once called 'wisdom'."[2]

Myth 2: Being old means being sick

Getting older does not mean one automatically becomes sickly. In fact, 85 per cent of people over the age of 65 have no real physical problems. Lack of exercise is what causes problems for the elderly, more than ageing itself. Most people in today's older generation grew up before exercise was emphasised, particularly for women. Result: many problems associated with old age, including muscular weakness, are really caused by lack of activity over a lifetime. But a number of studies have shown that older people who start to exercise can improve their strength and become more vigorous. As the current generation of exercisers grows older, they will remain fit for longer — and look far younger than previously.

Myth 3: Being old means being frail

It is possible, even probable, to live a long, full life and die without being physically incapacitated. People who stay active achieve what Los Angeles gerontologist Dr James Birren calls a "plateau of healthy aging".[3] This is a level that they maintain from their 60s until death. According to Dr Birren, many oldsters experience no extended period (more than a few months) of illness or disability in their entire lives.

Myth 4: You are old at age 65

Age 65 can no longer be considered old. In 1900, life expectancy was only in the mid–40s. Living to age 65 was rarer and 65 was truly "old" compared with almost everyone else living in the community. But life expectancy in Australia now is 78.8 years for women and 72.3 years for men. According to Dr Lydia Bronte, "the length of adulthood has almost doubled. Increased longevity has created what I call a second middle age between ages 50 and 75. Most of the people I interviewed [over age 65] do not think of themselves as old. They believe that as long as they can do what they want to do and enjoy it, there is no reason to stop." She claims that "virtually nobody is physically old at 65".[2]

The top 10 nations in life expectancy according to the World Health Organization are:

(1) Japan
(2) Australia
(3) France
(4) Sweden
(5) Spain
(6) Italy
(7) Greece
(8) Switzerland
(9) Monaco and
(10) Andorra.

The bottom 10 nations in life expectancy are:

(182) Ethiopia
(183) Mali
(184) Zimbabwe
(185) Rwanda
(186) Uganda
(187) Botswana
(188) Zambia
(189) Malawi
(190) Niger and
(191) Sierra Leone.

Myth 5: Older people always look forward to retirement

Dr Bronte argues that "most older people who work do so because they enjoy it. Even though retirement is supposed to be the pot of gold at the end of the rainbow, many older people claim they are happier working. Unfortunately, many older workers in America, and presumably in Australia too, have to be resourceful to keep working because so many companies 'cycle out' older employees. They cope by starting businesses ... moving to a company where retirement age is not mandated ... turning a hobby into a vocation."[2]

COULD I REALLY LIVE TO BE 400 YEARS OLD?

We looked at this in *The Odd Body*. Some scientists predict that in perhaps a decade or so humans will be able to halt or even reverse ageing through genetically engineered drugs. Theoretically, if ageing is overcome, then all the age-related diseases that eventually kill most of us such as stroke, heart disease and cancer will also be overcome. Finding drugs that will reverse ageing is the holy grail of longevity research.

Scientists know that ageing occurs on the genetic DNA level in the cell. Cells are born, live, die and are continually replaced by new cells throughout our lifetime. But the cells produced later in life are more often poorer in quality than the cells produced in youth. As these poorer cells build up in us, our bodies don't look as good and don't work as well. Enzymes produced by our bodies in large supply early in life "correct" cells from becoming poorer cells. But these enzymes aren't as plentiful later in life. Genetically engineered anti-ageing drugs will merely restore the enzymes. As a result, we could halt or reverse ageing. We could remain at whatever physical age we wished for as long as we wished. We would grow older chronologically, but stay the same physically. In theory, by modulating drugs so that ageing is reversed at the same rate as normal ageing occurs, we could, for instance, start taking the drugs at age 60 for 10 years, then at age 70 look and feel like we did at age 50. And should we continue for another 10 years, at age 80 we would look and feel like age 40. We could stop our body's clock or turn it backwards. Life expectancy would jump from about 80 years currently to about 400 years. Statistically, by age 400, a fatal accident will have caught up with you. That's the theory, anyway. Of course, if people are routinely living to 400 where are we going to put them? What will that do to family life? Careers? The world as we know it?

CAN A RECENTLY DEAD PERSON GET SUNBURN?

(Asked by Simon Stevens of Curl Curl, New South Wales)

If they could, then at least they wouldn't have to worry about dying of skin cancer! Getting sunburned is impossible if a person is dead. Sunburn is an inflammation of the skin and the tissues just beneath it caused by overexposure to the ultraviolet rays of the sun. The inflammatory response causes the redness and swelling of sunburn. This also involves the release of chemicals from cells that then attract other cells (blood cells such as lymphocytes) to the site of the sunburn. The inflammatory response requires energy. The cells have to have energy in order to produce and respond to chemical signals; they cannot do so if they are dead. In addition, even in a fair-skinned person it would take at least 15 minutes in very strong sunlight for skin to get sunburned. In 15 minutes the dead cells would not be able to function in a way that would produce a sunburned corpse.

IS IT POSSIBLE FOR RESEARCHERS TO ACCIDENTALLY CREATE A VIRUS THAT WIPES OUT ALL HUMANS?

The answer to this has to be "Yes." It sounds rather like the scenario of a B-grade science fiction movie, but for many years researchers around the world have been creating 1 deadly virus after another in the laboratory. If a human killer virus somehow got released into the open, we could have a disaster in the human community — from which we might not recover.

In January 2001, Australian researchers in Canberra experimented with viruses in an attempt to develop a way of controlling the mice that plague Australian agriculture. They created a genetically modified (GM) mousepox virus. However, this "super virus" was far more powerful and effective than anyone predicted. The virus killed off 100 per cent of those mice exposed to it that were not immunised against it. It also killed off half of the mice that *had* been immunised against it. The virus was created by simply inserting into a gene the coding for interleukin 4 (IL-4). IL-4 is a protein belonging to the cytokine family that develops and regulates thymus (T) helper-cells, mast cells, resting T-cells and activated B-cells. These are important in the immune system response. Mice are genetically rather similar to humans. Isn't that a reassuring thought? So what could happen to mice could quite easily happen to humans.[4]

In 2000, Japanese researchers in Kyoto, while experimenting with the AIDS virus and wanting to know "How does the AIDS virus grow?", stumbled onto a way to make the AIDS virus grow much, much faster. Making the AIDS virus grow faster is not exactly what we want, but we now know how to do it. In fact, making the AIDS virus slow down is 1 way of overcoming AIDS.[5, 6]

WHY WAS THE BLACK DEATH CALLED THE BLACK DEATH?

The Black Death was so called not because its victims usually turned black, either partially or completely, and either before or after death, nor because it was "black" in the sense of being evil. The Black Death was not called by that name until the 19th century; names for it at the time of various outbreaks included the Great Pestilence, the Great Death and the Sudden Death. In some countries it was named after the place where it was supposed to have originated. For example, it was called Moroccan Fever or Italian Death. In the early 16th century, Scandinavian scholars mistranslated the Latin term for plague, *atra mors*. *Atra* can mean "black", but in this context meant "terrible". Over the next couple of centuries the new name gradually became universal. In Britain it was used to differentiate the 1348–1350 plague that wiped out ⅓ of Europe from the 1665 Great Plague of London, so famously chronicled in Daniel Defoe's *Journal of the Plague Year*, published in 1665. As it happens, blackened flesh is not a symptom of the particular plague we usually call the Black Death. The Black Death came in three forms: bubonic, pneumonic and septicemic. All forms were caused by a bacterium called *Yersinia pestis*.

The bubonic plague was the most commonly seen form of the Black Death and the mortality rate was 30 to 75 per cent. The symptoms were enlarged and inflamed lymph nodes around armpits, neck and groin. The term bubonic refers to the characteristic bubo or enlarged lymphatic gland. Victims were subject to headaches, nausea, aching joints, fever of 38.3 to 40.5° Celsius (101 to 105° Fahrenheit), vomiting and a general feeling of illness. Symptoms took from 1 to 7 days to appear.

The pneumonic plague was the second most commonly seen form of the Black Death. Death was usually fairly quick. The mortality rate was 90 to 95 per cent. Pneumonic plague infected the lungs and the most telltale symptom was bloody sputum. Symptoms took from 1 to 7 days to appear.

The septicemic plague was the least commonly seen form of the Black Death. Death was very quick, often on the same day as symptoms appeared. The mortality rate was 100 per cent. Septicemic plague was an infection of the entire body. Symptoms were a high fever and skin that turned deep shades of purple due to disseminated intravascular coagulation of blood.

HOW LONG DOES IT TAKE MUSCLE TO DIE?

(Asked by Sam Gardner of Edmonton, Alberta, Canada)

Deaths are divided into 2 categories: somatic and cellular. Somatic death occurs before cellular death. Somatic death is the death of the whole body. It is the point at which the body can only be kept alive with the aid of machines. Cellular death occurs when individual cells can no longer metabolise. Different types of cells die at different rates. After somatic death, muscle cells cease to receive oxygen and die due to the lack of blood circulation. The muscle cells then use anaerobic respiration to produce energy (adenosine triphosphate, or ATP) from stored glycogen. One product of anaerobic respiration is lactic acid. Lactic acid cannot be removed from the cells, and their pH decreases, since the blood is no longer circulating. This low pH inhibits enzyme activity, which controls cellular metabolism. Therefore, the low pH causes a stop to muscle-cell metabolism.

The time that it takes for muscle-cell metabolism to cease depends in part on the environmental temperature at the time of death. The higher the temperature, the faster the chemical reactions occur that cause the cells to produce the lactic acid and thus speed up cell death. The rate at which muscle metabolism ceases also depends on the initial concentration of lactic acid in the cells at the time of death. For instance, if the circumstances of death invoked stress, the amount of lactic acid in the muscle cells at the time of death would be increased. This would result in a shorter time for metabolic activity to cease after death.[7]

WHAT IS RIGOR MORTIS?

(Asked by Ben Kind of Waverley, New South Wales)

Rigor mortis is the stiffening of muscles that occurs after death. The muscles stiffen and then loosen in a process that takes about 12 hours. However, temperature and a variety of other factors can alter this time framework. Coroners and forensic pathologists use rigor mortis as a major tool in determining the time of death. Rigor mortis is caused in part by the breakdown after death of the muscle cell's sarcoplasmic reticulum. The sarcoplasmic reticulum is composed of large amounts of calcium. When no longer moving and thus isolated after death, the calcium is released. This triggers muscle contraction. The calcium-release process produces what are called cross-bridges between specific proteins (myosin and actin) in cells. For muscle cells to relax, energy produced from respiration is required. After somatic death, muscle cells eventually lose their metabolic potential to produce energy and permanent cross-bridges are formed. These permanent cross-bridges cause the rigor mortis.

The time required to achieve rigor mortis varies with the type of species. Poultry may require only 1 to 2 hours, whereas cows are likely to need 20 to 24 hours. For humans in temperate regions, the following is generally used as a guide to estimate the time of death:

☞ If the environment is warm and the body is not stiff then the person is believed to have been dead no more than 3 hours.

☞ If the environment is warm and the body is stiff then the person is believed to have been dead between 3 and 8 hours.

☞ If the environment is cold and the body is stiff then the person is believed to have been dead between 8 and 36 hours.

☞ If the environment is cold and the body is not stiff then the person is believed to have been dead for more than 36 hours.[8]

WHAT IS LIVOR MORTIS?

(Asked by Ralph Parslow of Maitland, New South Wales)

Everyone has heard of rigor mortis, but less well known is livor mortis, the pooling of blood after death. Blood is a liquid. Gravity causes blood to find its lowest point within the corpse, since the heart is no longer pumping the blood throughout the body. In this sense blood in the body is just like juice in a can or water in a flood. Livor mortis also takes about 12 hours to complete. But just as with rigor mortis, many factors can change the time framework, so coroners and forensic pathologists also find livor mortis to be helpful in establishing the time of death.[9]

WHY DOES A PERSON WHO DROWNS SINK AND THEN FLOAT?

(Asked by Ralph Parslow of Maitland, New South Wales)

A drowning person will often suck water into their lungs, which will reduce buoyancy, causing the person to sink. Micro-organisms will begin the process of decay once the person is dead. They release gases that can cause the buoyancy of the body to increase, thus the body tends to float. Only 10 per cent of drowning victims die from lack of air. The other 90 per cent panic and inhale water, which kills them faster than holding their breath would. Even if the lungs are dry and full of air, people die after 4 to 10 minutes of not breathing. The average living person has a specific gravity very close to that of water. Heavy clothing may increase the time a corpse spends underwater, but will not usually prevent it from eventually rising.

WHY ARE GERMS SO BAD FOR YOU?

(Asked by Scotty Harper of Toronto, Ontario, Canada)

Germs are not necessarily bad for you. A germ is any small, rudimentary piece of living organism capable of reproducing. A germ can be a seed, a pod or a cell. That is why, for instance, the small bit of living protoplasm in a wheat kernel is called wheat germ. However, various small bacteria and viruses have popularly become known as germs despite the fact that the meaning of the word is more general. In fact, there are 4 major types of germs: bacteria, viruses, fungi and protozoa. They can invade plants, animals and humans. Once germs invade the human body, they consume nutrients and energy while producing toxins. These toxins can cause infections, fevers, coughing, sneezing, rashes, diarrhoea and vomiting. Sometimes germs can make you sick and are therefore bad for you — but not always.[10]

DO MOST PEOPLE DIE IF STRUCK BY LIGHTNING?

(Asked by Jenny Jordon of Coral Gables, Florida, USA)

Contrary to what is often thought, most people survive being struck by lightning. One estimate claims that 70 to 80 per cent of people struck by lightning recover. However, they often have hearing loss, eye damage or muscular problems. Surprisingly, the burns received are not usually severe, because the lightning strike itself is so fast (about 100,000 miles per hour) that flesh does not have time to burn.[11]

WHERE ARE YOU MOST LIKELY TO BE STRUCK BY LIGHTNING?

(Asked by Jenny Jordon of Coral Gables, Florida, USA)

According to Professor Stephen A. Nelson,[12] 26 per cent of lightning deaths occur in open ground or sports fields, 15 per cent under trees, 12 per cent on boats and in water-related activities, 5 per cent on golf courses, 6 per cent on tractors and heavy road equipment and 1 per cent while talking on the telephone. The remaining 35 per cent of deaths are classified as "unspecified". Home deaths from lightning can occur as a result of a side flash — when someone is standing near a chimney, a fireplace, plumbing, electrical appliances, the television or, again, the telephone.

You're most likely to be struck by lightning in the head, shoulders or feet.

Queensland is the Australian state with the most lightning deaths; Florida is the US state with the most lightning deaths.

Lightning kills more people annually in the US than do tornadoes, hurricanes or any other natural weather force.

Lightning can travel from clouds to the ground, but also from the ground to the clouds and within clouds themselves.

The human body itself generates lightning. When you walk across a carpet during a low-humidity day and then touch a metal object such as a doorknob, a static electrical charge is built up and then discharges as a tiny bolt of lightning. How shocking! Hope you get a charge out of this!

***

DID A WOMAN ONCE TURN TO SOAP?

(Asked by Will Burridge of Liverpool, New South Wales)

She did! You can see her. And she's not the only one. It is a process called human saponification, which literally means "turning into soap".

The famous Soap Woman is on display at the Mutter Museum in Philadelphia. Her story is truly bizarre. Some time in the 19th century, a woman who was rather fat died of yellow fever. After her burial in a Philadelphia cemetery, the fat of her body turned into adipocere, a fatty wax substance that feels similar to lye soap. She became saponified in this way when her body fat reacted to the combination of chemicals in the soil. She has been on display at the Mutter Museum since 1874, when Dr Joseph Leidy, a prominent University of Pennsylvania anatomist, donated her body. According to Dr Leidy, the Soap Woman was named Ellen Bogen, who died in 1792 near Fourth and Race Streets in Philadelphia. Her body was uncovered by workmen removing bodies from an old burial yard. In 1942, the Museum curator, Joseph McFarland, determined that the Soap Woman actually died much later. He discovered that there was no yellow fever outbreak in 1792 and there was no census record of a woman named Ellen Bogen at that time. Eight pins and two 4-hole buttons on the clothing she was wearing when buried were dated as being from the early 19th century. This helped museum staff establish the approximate date of her death as some time in the 1830s or 1840s. In 1987, X-rays of her body allowed researchers to determine that she was about 40 years old when she died. In

2001, museum staff performed a CT scan on the Soap Woman to learn more about the process that led to her strange fate. The Mutter Museum of the Philadelphia College of Physicians was founded in 1849. It was formerly only open to medical professionals and scholars, but it is now open to the public. An accompanying display shows an X-ray cross-section and tells the story of the Soap Woman. A Soap Man who was buried alongside the Soap Woman is sometimes displayed at the Smithsonian Institution in Washington, DC.[13]

HOW LONG DOES IT TAKE A BODY TO DECOMPOSE?

(Asked by Jason Long of Wollstonecraft, New South Wales)

The rate of decomposition depends on the state of the body and the environment in which it is kept. For example, an unembalmed body left on the ground in a temperate climate and exposed to open air will decompose in 1 day to the same extent as a body left in water will decompose in 1 week or a body underground will decompose in 1 month. The temperature of the air, ground or water affects the rate of decomposition. Heat speeds it up, cold slows it down. Decomposition is slower in sea water than in fresh water. It is also slower in running water than in stagnant water. The lower the temperature of the water, the slower the decay of the body. An unembalmed body buried 1.8 metres (6 feet) deep in ordinary soil without a coffin takes 10 to 12 years to decompose down to the skeleton. It takes about 300 years for the bones to fully decompose under those circumstances. Bodies left on the surface in a temperate climate could decompose within 4 to 6 weeks. The higher rate is due to insects and animals that eat away at the bodies. Embalming slows down the decay of the body by a factor of 2 to 10, but it doesn't alter the rate of bone decay.[14, 15]

WHY DO SOME PEOPLE DIE FROM EATING PEANUTS?

(Asked by B. Hawkins of Napa, California, USA)

People are allergic to the proteins found in peanuts. Both fresh and roasted peanuts can cause an allergic reaction, since the proteins are not destroyed in cooking. Half the adults and more than half the children who have an allergy to peanuts are also allergic to other nuts such as walnuts, cashews or Brazil nuts.

Symptoms of peanut allergy vary from person to person. Most people develop symptoms immediately after coming into contact with peanuts. Minor symptoms can include a mild rash, tingling mouth and tongue or stomach ache. Serious symptoms (anaphylactic reaction) can include facial swelling, difficulty breathing, collapse, coma and even death. The degree of allergic reaction also varies from person to person. Some people are so sensitive that they can react after merely touching someone who has eaten peanuts or being in the same room with peanuts. According to the Institute for Child Health at the famed Great Ormond Street Hospital in London, "Anyone can develop an allergy to peanuts, but it is more common in people who have other atopic conditions [inherited allergies], like eczema, asthma or hay fever, or who have immediate family with these conditions." The Institute adds that "There is some discussion as to whether eating peanuts and peanut products in pregnancy increases the risk of a child developing peanut allergy. The results are not conclusive, but the [UK] Department of Health recommends that if a member of the immediate family has an atopic condition, the mother should avoid peanuts and peanut products during pregnancy and breastfeeding. If you think your child is at risk, then avoid peanuts or peanut products until he or she is at least three years old."[16]

WHY DO RIGHT-HANDERS LIVE LONGER THAN LEFT-HANDERS?

(Asked by G. Rohrmann of Tampa Bay, Florida, USA)

There is debate among scientists as to whether right-handers live longer than left-handers and if so, why. The bulk of the evidence suggests that right-handers do indeed live longer than left-handers. In 1988, Dr Diane Halpern and Dr Stanley Coren presented findings that left-handers have shorter life spans. They argue that left-handedness could indicate birth-stress-related neuropathy, developmental delays and even immune system deficiencies.[17] In 1989, Dr W.P. London argued that lower life expectancy in left-handers is not surprising, since they also have higher rates of alcoholism, smoking, breast cancer and several serious neurological and immune disorders.[18] But that same year, Dr Max Anderson reported that left-handers had *longer* life spans.[19]

Then, in 1992, Dr Charles Graham and colleagues at the Arkansas Children's Hospital in Little Rock concluded that left-handers do indeed live shorter lives. This is because in an admittedly right-handed world, left-handed children have more accidents — some of which are fatal. Equipment and mechanical devices are not designed with the 11 per cent left-handed minority in mind — often with disastrous consequences. But according to Dr Clare Porac, "Left-handedness is not necessarily the kiss of death." She notes, "Although the percentage of left-handed people among those over age 60 is lower than in the rest of the population, there is no indication that left-handedness leads to an early demise. Rather, a complex combination of factors combine so that fewer of the old and oldest old report left-handedness." She adds that one of these factors is that 80 per cent of left-handers over 75 years of age were pressured to change handedness preference when they were children — with mixed results. Dr Porac, along with Dr Ingrid Friesen, presented these findings in 2000.[10] Later that year, a study by a team of Danish researchers from Aarhus University found no relationship between handedness and mortality. It looked at all twins born in Denmark in 1910 and 1911, most of whom were dead.[21] There has been little research on this question since 2000.[22]

IF BOTH MY HANDS WERE SEVERED AND MY LEFT HAND WERE SEWN ONTO MY RIGHT ARM (AND VICE VERSA) WOULD I STILL BE RIGHT-HANDED OR LEFT-HANDED?

(Asked by D. Sonius of Chipping Norton, New South Wales)

Handedness is governed by the brain and not by the hands themselves. Therefore you would still be whatever you were before, right-handed, left-handed or even ambidextrous. But if your left hand were sewn onto your right arm and your right hand were sewn onto your left arm, your thumb on each hand would be your outermost digit and not your innermost digit as it ought to be. You would be weirdly unique among humans, and certainly entitled to sue your surgeon for malpractice.[22]

DO CANNIBALS STILL EXIST?

Cannibalism (anthropophagy) has probably existed since humans first became humans. Neanderthals practised it some 200,000 years ago, according to a team of French and American archaeologists. In 1999, they discovered evidence of Neanderthal cannibalism in a cave in south-eastern France. Regions of the world where cannibalism was probably practised include the West Indies (among the Caribs, from whom we get the name "cannibalism"), west and central Africa, Melanesia, Polynesia (e.g., among the Fijians), Australia, New Zealand, Central America (e.g., among the Aztecs of Mexico) and North America. One site in south-western Colorado indicates that humans were butchered and eaten by other humans there as late as 850 years ago. Dr Richard Marlar and colleagues from the University of Colorado School of Medicine have found thousands of human bones in a number of sites suggesting cannibalism.

Cannibalism occurs for many reasons. Sometimes food is scarce. More often the reason has to do with religion, ritual, magic, warfare, crime and punishment, feuding and revenge, or a combination of these. A feared enemy slain in warfare may be ritually eaten by his slayer in the belief that by consuming his body, his courage, strength and spiritual power may be transferred to the victor. This is called endocannibalism.

Instances of cannibalism by groups or individuals occur from time to time. As for groups, Napoleon's starving army resorted to cannibalism as they retreated from Moscow in 1812. Survivors of the frigate *Medusse*, drifting in lifeboats, were forced to eat those too weak to survive in 1816. So were members of the Donner Party, snowbound and trapped in the Sierra Nevada Mountains in the winter of 1846 to 1847. Survivors of a plane crash in the Andes ate dead passengers to stay alive in 1972.

As for individuals, Fritz Haarman, the "Hanover Vampire", is infamous. In Germany in 1924, he murdered 27 boys and made them into sausages that he sold to the unsuspecting. Jeffrey Dahmer in Michigan is also infamous, as is Armin Meiwes in Germany, who murdered and partially ate a person he met in an Internet chatroom. There persists the occasional report of cannibalism being practised in remote areas of Papua New Guinea and central Africa. But evidence for this is rare — medium rare.

ARE HUMANS NORMALLY WARLIKE OR PEACEFUL?

Evidence suggests that a state of war rather than peace is the "normal" state of humanity. Indeed, some anthropologists argue that nearly every society, ancient or modern, has gone to war fairly regularly throughout the course of its existence. From the smallest tribes to the largest nations, we humans have been at war with each other almost nonstop for 5000 years (from at least 3200 BC to the present). In the last century or so, there has been little improvement on this sorry record. The world has witnessed more than 150 wars since the end of World War II. World War II was the bloodiest in human history. Forty-five million people perished.

A longtime researcher of the origins of war is anthropologist Dr Keith Otterbein. He theorises that there are 3 conditions at the root of the warfare motivation. First, there are "ultimate" or "survival and hate" reasons. These are immediate causes for conflict, such as competition for scarce food, water or mates, or intense divisions and antagonisms between groups of related men. Second, there are "proximate" or "elbow room" reasons. These are secondary causes for conflict, such as a people's (or perhaps only a leader's) desire to conquer land even if it is not really needed for survival, a desire by the military to use weapons that the military has in its arsenal, and so on. Third, there are "war breeding war" reasons. That is, the results of earlier wars teach humans that war may be profitable as well as costly. The hope of winning a war encourages more war — for the gain of resources, prestige, wealth and power.[23]

Besides Otterbein's theory, there are many other theories as to why a state of war is the "normal" state of humanity. Yet although warfare seems to be much more "normal" than peace, anthropologists note a few societies where warfare is completely unheard of. Nice to know there are places where war is "abnormal".

ARE HUMANS MORE WARLIKE NOW?

(Asked by F. Billington of Storrs, Connecticut, USA)

This is like asking "Are humans basically good or basically evil?" and being answered "Humans are basically good, but proving they're basically evil is easier."

Our world seems so hostile. Our cities and neighbourhoods seem more violent every day. TV and other media seem to be more vicious. One could easily be forgiven for thinking that the world is becoming more and more hopelessly warlike. But believe it or not, human warfare is actually declining. Over the 20th century, combat deaths exceeded 100 million and over 170 million people died at the hands of governments. This was due to the enormous growth of war technology which still continues at an alarming rate. However, as a ratio of total population, adult war deaths have plummeted, especially in modern democracies. As anthropologist Lawrence H. Keeley argues: "Archaeological evidence tells us that with very few exceptions, most foraging societies and early agricultural civilizations were at war almost incessantly — or at least far more frequently than modern societies today. Whereas among foraging and early agricultural societies an average of 25 per cent of adults would die by warfare, this figure fell to less than 10 per cent in early modern societies (about 200 years ago)."[24] According to Lloyd deMause, an average of less than 1 per cent of adult deaths among modern democratic nations today are war casualties.[25] This should make us pause, and maybe generate a little hope. It seems that even though we still perceive that we live in a warlike world, the average citizen of the average modern democracy knows more peace now than our ancestors could ever have imagined.

WHAT HAPPENS WHEN YOU BREATHE YOUR LAST?

Cheyne–Stokes breathing is the type of breathing we engage in when we are about to die. When doctors and nurses observe this form of respiration in a patient, they know that the patient may have only a few days, hours or perhaps minutes to live. Almost certainly the patient is experiencing congestive heart failure. Probably the patient is in a coma, or nearly so. Possibly the patient has severe brain damage. These associated brain problems are called Cheyne–Stokes psychosis and also involve anxiety and restlessness. It is not a comforting sight. The brain is struggling for consciousness and the body is struggling for air, and both are losing the battle.

In Cheyne–Stokes breathing there is a pattern of gradual increase in the depth of respiration. This is followed by a progressive decrease in the depth of respiration, resulting in apnoea-like breathing. To imagine this, take a deep breath and hold it for as long as you can and then let it out all at once. Thereupon you take another deep breath and do the same. In Cheyne–Stokes breathing it is as if the entire body is focused solely upon maintaining breathing, and its single motivation is to keep breathing so it will stay alive for one breath longer. Cheyne–Stokes breathing continues as the body desperately fights on.

Humans are hard-wired to survive and to fight death until the last. If we did not fight so valiantly for life whenever faced with the challenge, we would not have survived as a species for so long. When the exhausted body can make no further effort, the patient breathes their last and death ensues. Rarely does the patient emerge from Cheyne–Stokes breathing or the associated Cheyne–Stokes psychosis, but it has been known to occur. In death as in life, miracles do happen.

WHAT HAPPENS WHEN YOU ARE EXECUTED BY HANGING?

(Asked by R. Smith of Hamilton, Bermuda)

Who do you mean by "you"? Hopefully none of us will ever have to face this horrible form of death. A widespread myth about hanging as a form of execution is that death is caused by strangulation. Instead, death usually occurs by a massive trauma to the upper part of the spinal cord from the major fracture, or massive dislocation, of cervical vertebrae. It's a broken neck if you don't define "neck" too precisely.

For centuries, hanging was one of the most common methods of execution. Until the late 1700s, when people began to develop more humane forms of killing, the person was simply pushed off a short height with a rope casually placed around the neck. A slow death from strangulation was the usual, often intended, result. Hanging was widely practised. For example, in England at the time of the American Revolution, it was often a hanging offence to steal a bakery item or poach small game from a lord's property. Interestingly, the rape of a child was not a hanging offence. Even into the 1800s, paedophilia was rampant and the age of sexual consent was 6. For many centuries the sexual abuse of a young child was less punishable than the theft of a loaf of bread or a rabbit.

The method of hanging was eventually modified. Various and higher types of drops were developed to cause a broken neck instead of strangulation. The proper placement of the knot under the left ear meant that during the fall the head was jerked with sufficient force to break the neck. The weight of the victim and the quality of the rope were included in calculations. Even today, hanging advocates claim that hanging is a sudden and usually instantaneous form of death. They argue that, although not a pretty sight for witnesses (what form of execution is?), hanging is speedier than death by electrocution, lethal gas, firing squad or guillotine.[26]

WHY IS EVERYONE AFRAID OF LEPROSY?

(Asked by S. Vance of Orinda, California, USA)

People should not be afraid of leprosy (Hanson's disease). It isn't the horrible, ugly, highly contagious, incurable and disfiguring disease many people still think it is. In fact, leprosy is actually 1 of the least contagious of all infectious diseases. It is less infectious than the common cold.

There are 2 types of leprosy: the acute type is a more mild tuberculosis form and is not very infectious at all; while the lepromatic type is somewhat more persistent but only slightly more infectious. But both types of leprosy have been completely curable since 1950, and with early detection and treatment leprosy leaves no disfigurement at all. According to the World Health Organization, "Today, diagnosis and treatment of leprosy easy."

Leprosy is the world's oldest disease, with documented cases dating back to 1350 BC. Biblical scholars contend that Bible references to lepers do not relate to the specific disease we now know today as leprosy. Instead, they probably refer to people who were morally and physically unclean or who had a variety of disfiguring and deforming diseases. Nevertheless, negative attitudes towards leprosy have continued throughout history. Leprosy was often "treated" through organised segregation in leper colonies. People knew very little about disease until the middle of the 19th century — and even less about treatment.

Interestingly, the precise way in which leprosy is acquired is not fully understood even today. It is a tropical disease and a disease of poverty. Those living outside the 30th parallels of latitude avoid the disease. Deformities are not as severe or common as are often imagined. The classic 1959 movie *Ben-*

Hur depicted leprosy in a highly exaggerated manner. Except to facilitate treatment, it is now recognised that the isolation of leprosy patients is unnecessary. According to the World Health Organization, the number of leprosy patients under treatment in the world was around 460,000 at the end of 2003. In March 2005, the World Health Organization stated that "during the past two years, the global number of new cases detected continued to decrease dramatically (a reduction of about 20 per cent per year)".[27] Leprosy has been virtually eliminated everywhere in the world except in parts of Angola, Brazil, India, Madagascar, Mozambique, Nepal and Tanzania.[28]

About 90 per cent of all humans who have ever lived are dead.

Currently, only 1 person in 2 billion lives to be 116 or older.

On average, every year in the US, 100 people choke to death on ballpoint pens.

The longest-living authenticated female twins were Kim Narita and Gin Kanie of Japan. They were born on 1 August 1892. Kim died of heart failure in January 2000 at 107 years of age.

The longest-living triplets were Faith, Hope and Charity, who were born in Elm Mott, Texas, on 18 May 1899. Faith was the first to die, at age 95, in October 1994.

The world's longest-living quadruplets were the Ottman siblings, Adolf, Anne-Marie, Emma and Elisabeth. They were born on 5 May 1912. Adolf died first, at age 79, in March 1992.

The man with the oldest known individual ancestor is British professor Adrian Targett. Tests matched Professor Targett's DNA with that of a 9000-year-old skeleton found in Cheddar, England.

Does lightning strike the same person twice? Roy Sullivan, a park ranger from the US state of Virginia, has been struck 7 times by lightning. The strikes occurred between 1942 and 1977. Having survived these strikes, "the human lightning rod" took his own life in 1983, reputedly over a failed romance.

Strictly speaking, nobody is "buried" in Grant's Tomb. President Ulysses S. Grant and Mrs Grant are "entombed" in New York City. A body is "buried" only when it is placed in the ground and covered with dirt.

The average human body contains enough sulphur to kill all the fleas on the average dog.

WHAT IS MOTOR NEURONE DISEASE?

It's a mystery disease. We know very little about it. It has no cure. About 1400 Australians suffer from it. It kills 1

Australian every day. It's killed perhaps the greatest all-round athlete the world has ever known. It's now slowly destroying the body of perhaps the greatest scientist the world has ever known.

Motor neurone disease (MND) is actually a group of related diseases that affect the motor neurones in the brain and spinal cord. Motor neurones are the nerve cells along which the brain sends instructions to the muscles in the form of electrical impulses. Degeneration of the motor neurones leads to weakness and wasting of muscles. This generally occurs in arms or legs initially, with some groups of muscles being affected more than others. Some people may develop weakness and wasting in the muscles of the face and throat. This causes problems with speech, and difficulty chewing and swallowing. Interestingly, MND does not affect touch, taste, sight, smell or hearing, nor does it directly affect bladder, bowel or sexual functioning. In the vast majority of cases, the intellect also remains unaffected.

The early symptoms of MND are mild. They include problems with walking, weakness in the hands, difficulties in holding objects, difficulty in keeping balance, slurring of speech and problems with swallowing. Symptoms get worse as more and more muscles shut down due to further deterioration of their activating neurones. Usually MND affects a single limb at first and later becomes distributed throughout the body. There is often an unusual twitching (fasciculation) that can be seen when looking at the muscle. When this is visible, it is a sign that the lower motor neurones of the brain have been damaged. Some sufferers with MND experience pain, but most, fortunately, do not. Leg cramps seem to be the pain most often reported.

The cause of MND is unknown. Theories include viral agents, environmental toxin and chemical exposure, immune system compromise, rogue protein destruction and premature ageing of motor neurons, but all of these are open to speculation. Heredity is thought to be a factor in MND, but perhaps only a minor factor. There is a genetic link in only 10 per cent of MND cases. In 1993, a specific gene on chromosome 21 was found to be mutated in 15 to 20 per cent of MND sufferers.

Diseases related to MND are progressive muscular atrophy (PMA), progressive bulbar palsy (PBP) and primary lateral sclerosis (PLA). MND goes by other names too, such as its US name, amyotrophic lateral sclerosis (ALS), Lou Gehrig's disease and, now, Stephen Hawking's disease. Lou Gehrig was a famous US baseball player early in the last century. He played baseball for 17 years, from 1923 to 1939, yet he was regarded as such an all-round athlete that he would have been a champion in any sport. The first baseman for the famous New York Yankees, Gehrig was known for his smooth, powerful and mighty swing. So solid and strong was his body that, despite injury and illness, Gehrig played 2130 consecutive games. His streak remained unmatched for 56 years. At the beginning of the 1939 season, Gehrig noticed that he became dizzy when he attempted to stand up, could not hit or throw with as much power and had a far slower reaction time. Tests revealed he had MND. His playing career was over and his days were numbered. On 4 July, Gehrig was given an appreciation day at Yankee Stadium. In 1 of the most famous scenes in sports history, a tearful Gehrig, knowing he was doomed, thanked the crowd from home plate for their goodwill shown throughout his 17 seasons. He said, "Fans, for the past two weeks you have been reading about the bad break I got. Yet today, I consider myself the luckiest man on the face of the earth." Tragically, he was dead within 2 years.

MND does not affect the mind. Stephen Hawking attests to this. The Cambridge University physicist and author of *A Brief History of Time*[29] and other books has been battling MND since 1963. This is truly incredible. His active brain spins theoretical ideas better than any spider spins a web. Now confined to a wheelchair, Hawking was not always so debilitated. Indeed, he grew up as a healthy young man. Hawking recalls that although he was never particularly good at sport, he did participate in inter-college rowing while studying at Oxford. In his third year, he noticed that he often fell over for no apparent reason. A year later, just after his 21st birthday, Hawking was diagnosed with MND. He says he was shocked to learn he had an incurable disease, but he was determined to fight it. He found strength by listening to Wagner and turning his attention to academic pursuits. He eventually finished his PhD and now holds the position once held by Sir Isaac Newton, Lucasian Professor of Mathematics at Cambridge. He has survived with MND for more than ⅔ of his life. MND is not a death sentence. There is definitely hope.[30]

WHAT KILLED DUDLEY MOORE?

Dudley Moore was born in 1935 and died in 2002. He first gained international fame in the 1960s as half of a comedy team with Peter Cook. He later achieved even greater fame as the diminutive star of such films as *10*, *Arthur*, the remake of *Unfaithfully Yours* and many others. Moore died of a rare, little known and 100 per cent fatal disease called progressive supranuclear palsy (PSP). In the last years of his career, he acquired the reputation for unreliability. He figured in some unflattering news stories involving driving problems, marital problems and spousal abuse. These were first believed to be alcohol-related, but the PSP diagnosis could also explain these behaviours. PSP slowly robs the body of muscular control. It is related to other slow neurodegenerative brain diseases such as Parkinson's disease, Alzheimer's disease and motor neurone disease. Actor Michael J. Fox and former US Attorney General Janet Reno suffer from Parkinson's disease, as did Pope John Paul II. Former US President Ronald Reagan suffers from Alzheimer's disease.

Medical science is desperately trying to understand and cure PSP, but such efforts have come too late to save Moore. The first symptom Moore noticed was his inability to keep his balance. Being a concert pianist, he also noticed difficulty in the movements of his fingers while playing. Following heart surgery in September 1997, Moore's symptoms became worse. After more than 5 years of suffering from symptoms he thought were related to his alcohol intake, Moore sought help in January 1998. He could no longer play his beloved piano and had difficulty speaking, standing, sitting upright in a chair and walking more than short distances. It took the finest experts in the world 13 months to finally diagnose PSP.

Outside the medical profession, most people know little or nothing about PSP. It is also called Steele–Richardson–Olszewski disease, after the 3 doctors who discovered it. PSP is indeed rare; prior to 1999 it was believed that it occurred in only about 1 in 100,000 people. However, a survey of PSP in the UK that year found that actually about 6.4 people per 100,000 suffer from it.[31] According to Dr Stephen Reich, the doctor who treated Dudley Moore at the Johns Hopkins Medical School in Baltimore, "because the earliest signs of PSP are often quite subtle, and sometimes take several years to evolve, the diagnosis is typically delayed. At the onset of symptoms, patients are often misdiagnosed as having Parkinson's, Alzheimer's disease, strokes, alcohol intoxication, or even inner-ear diseases."[32] Early symptoms of PSP include falling, difficulty walking, balance disturbances and slow movements generally. PSP sufferers may also have difficulty swallowing; indeed, this is how most PSP victims die. They eventually lose the ability to swallow (dysphagia) and usually choke to death (dysarthria). According to Dr H.R. Morris and colleagues, in the case of PSP, the brain's "neuronal degeneration is associated with the deposition of hyperphosphorylated tau protein as neurofibrillary tangles".[33] In layman's terms: just as the electrical wiring of a house can short out, the brain's electrical circuitry of nerves shorts out due to clogging from the build-up of a protein residue. The more clogging that occurs, the more difficult it becomes for the brain's messages to reach the body's muscles — and the more difficult it becomes to control movement. The big mystery for medical science is what triggers the protein residue to build up and how the process can be halted and reversed.

There is some hope, however. The cause of PSP may be slowly revealing itself. For example, a team of Spanish researchers led by Dr M. Ezquerra have reported the isolation of what they think is the gene abnormality that triggers PSP.[34] Medical science has also learned that the protein residue clogging the brain in PSP is identical to the residue clogging the brains of Parkinson's, Alzheimer's and motor neurone disease sufferers. Perhaps a cure for 1 disease may lead to cures for the others.

Almost until the day he died, Dudley Moore tried to make light of his tragic condition. In September 1999, when he announced he had PSP, he said: "I understand that one person in 100,000 suffers from this disease, and I am also aware that there are 100,000 members of my union, the Screen Actors Guild (SAG) who are working every day. I think, therefore, it is in some way considerate of me that I have taken on the disease for myself, thus protecting the remaining 99,999 SAG members from this fate." Always looking for humour, rest in peace, Dudley Moore.[35, 36]

CAN OWNING A PET MAKE YOU LIVE LONGER?

(Asked by James Raikar of Engadine, New South Wales)

There is plenty of evidence that your pet may help you live longer. In 1 of several studies, researchers gave dogs to 75 elderly people who were at risk of dying earlier than normal. They had serious health problems, lived alone and had low incomes. The elderly people thrived with their new pets, compared with those in a control group who were not given pets. According to the head of the study, Dr Daniel Lago, "People who were sick and who might have been expected to die simply did not die." He adds, "In our study, pets cut the risk of illness and death by about 50 per cent." Dr Lago gives 2 reasons for this. First, "Pet owners who lived alone no longer felt alone. They had companionship, which reduces anxiety and stress — real killers." Second, "I think probably the most important need being met is feeling useful — having something that is dependent on you. That makes you more responsible. You have to keep going."[37, 38]

Besides companionship and dependency, there are many other reasons pet ownership boosts health. Pets offer unconditional love and are ideal recipients of *our* unconditional love. We humans are social and emotional animals — we need to both give and receive love. Pets allow you to relax around them in a stress-free manner. This is not always possible when you're around people. Pets can physically stimulate you. For example, when Rover needs to go for a walk, you *have* to take him; if it were just yourself you might not bother. A study by the US Department of Health concluded that pets increase the survival rate of heart attack victims: 28 per cent of heart patients with pets survived serious heart attacks, while only 6 per cent of pet-less patients survived such attacks. In another study, the LDL cholesterol (bad cholesterol) levels of pet owners was 2 per cent lower than the bad cholesterol levels of pet-less people. This doesn't seem like much, but this reduction reduces the risk of heart attack by 4 per cent. Yet another study shows that owning a pet can reduce blood pressure just as much as eating a low-salt diet or drinking less alcohol. Still another study has found that nursing homes that use "companion animal therapy" have experienced a significant reduction in the use of prescription drugs. This has resulted in a reduction in the overall costs of caring for seniors as well. A US survey of

Medicare patients has shown that 40 per cent of elderly patients sought the services of a doctor much less often than those without a pet.[39, 40]

DO DEAD PEOPLE REALLY TWITCH?

(Asked by James Marr of Bexley, New South Wales)

Indeed, the dead do exhibit some motion for a few hours after death, until rigor mortis sets in. Even after that, some movement can be seen, often much to the shock of onlookers.

Every motion we make is a result of electrochemical action in our nerves. Chemicals are stored in our nerve endings and are released in the early post-mortem period. They can sometimes cause contractile muscles to move, resulting in a brief jerking motion in a limb. Dr Erin Cram describes 1 rather grim tale of a police officer who was killed in a shoot-out. For 3 hours his fellow officers tried to reach him, seeing his legs twitching and 1 of his arms moving as if he was trying to wave for help. When finally they were able to get to him and drag him to "safety", they found that he had been shot twice in the head and had died instantly. A twitch of a leg or a brief clench of the fingers is usually all that is seen. It is simply a chemical reaction, where muscle tissue is responding to a stimulus that it receives, even though that stimulus is not coming from a functioning brain.

Contrary to many an urban legend, a corpse will not sit up on the examining table or start talking at the morgue. However, in the centuries before embalming was practised, decomposition within the tissue of a corpse would cause gases to build up and this could lead to the occasional shifting of an arm or leg or the turning of the head. Sometimes there could even be an audible groan from escaping air moving past the vocal cords.

WHAT IS THE "BASKERVILLE EFFECT"?

This phrase refers to the phenomenon of scaring yourself to death. In the legendary Sherlock Holmes story *The Hound of the Baskervilles*, by Sir Arthur Conan Doyle, Sir Charles Baskerville dies from a heart attack brought on by extreme psychological stress. According to US sociologist, Dr David Phillips, people can indeed be scared to death, in fact as well as in fiction. Dr Phillips maintains that there is strong evidence linking extreme psychological stress and fatal heart attacks. He claims. "I have often wondered if people could indeed die by fright and if so, how this could be investigated quantitatively. I recalled that in *The Hound of the Baskervilles*, Sir Charles Baskerville dies of a fatal heart attack, apparently because he is frightened to death by the hound. Since Arthur Conan Doyle was a physician as well as an author, I wondered if his story was based on medical intuition or literary license, i.e. were fatal heart attacks and stress linked in fact as well as fiction?"

Although numerous laboratory studies have shown cardiovascular changes following psychological stress, for obvious ethical reasons only non-fatal stressors can be studied in the laboratory. Thus one may not be able to generalise beyond these mild stressors to determine if, in the real world, fatal heart attacks are precipitated by extreme stress. According to Dr Phillips, "The challenge was to find a way of testing this hypothesis that would circumvent the ethical problems of the laboratory experiment and yet retain some of its vigor. The best solution seemed to be to use a natural experiment — a real life event that met certain criteria. First, this real life event had to have distressing psychological effects on one segment of the population but not on others. In addition, it shouldn't actually be a dangerous event — it should only be perceived as such. Lastly, this event should not be linked to any changes in the quality of medical services."

It was not easy for Dr Phillips to find such an event that met all of these criteria. Nevertheless, he did. The real-life event was connected with a Chinese and Japanese superstition. In Mandarin, Cantonese and Japanese, the words for "death" and "four" are pronounced almost identically. As a consequence, the number 4 evokes discomfort and apprehension in many Chinese and Japanese people. Because of this, the number 4 is avoided and omitted in some Chinese and Japanese street, floor and room numberings, restaurants and telephone numbers. In addition, China's air force avoids the number 4 in designating its military aircraft because of the superstitious association between 4 and death. Interestingly, Dr Phillips found that cardiac deaths peak on the fourth of the month for Americans of Chinese and Japanese ancestry. Yet this is not a pattern seen among Americans of European ancestry. Dr Phillips studied computerised US death certificates to examine more than 200,000 Chinese and Japanese deaths along with 47 million white deaths, from 1973 to 1998. According to Dr Phillips, "Conan Doyle suggests that Sir Charles Baskerville was particularly susceptible to a stress-induced heart attack because he had a chronic heart condition. If Doyle's medical intuition was correct, deaths from chronic heart disease should display a particularly large fourth-day peak. Sir Charles Baskerville's superstitious fear of an avenging spectral hound was shared and reinforced by his neighbours. Similarly, Chinese and Japanese superstitious fears are likely to be stronger where they are reinforced by large Chinese and Japanese populations."

The evidence presented by Dr Phillips supports both these expectations. For Americans of Chinese and Japanese descent, there are 13 per cent more cardiac deaths than expected on the fourth of the month. This fourth-day increase is larger (27 per cent above expected) in California, where Chinese and Japanese populations are concentrated. Dr Phillips tested 9 alternative, non-psychosomatic explanations for these findings, including the possibility that on the fourth Chinese and Japanese might change diets, increase alcohol consumption, refuse medicines or overstrain themselves. Dr Phillips concluded that the data suggests a link between psychological stress and heart attacks. He says, "Our findings are consistent with the existence of psychosomatic processes, with the scientific literature, and with a famous non-scientific story. The 'Baskerville effect' seems to exist both in fact and in fiction, and suggests that Conan Doyle was not only a great writer, but a remarkably intuitive physician as well."[42]

Each year 3 people die in Australia while using their tongue to test if a 9-volt battery still works.[43]

In 1998 142 people in Australia were injured by not removing all the pins from new shirts.[43]

Each year 58 Australians are injured by using sharp knives instead of screwdrivers.[43]

Between 1996 and 2000, 31 people in Australia died by watering their Christmas tree while the fairy lights were plugged in.[43]

In the 3 years prior to 2000 19 Australians died by eating Christmas decorations they believed were made of chocolate.[43]

Hospitals in Australia reported 4 broken arms in 1999 after cracker-pulling incidents.[43]

⧝ ⧜

Between 1997 and 2000, 101 Australians had to have broken parts of plastic toys pulled out of the soles of their feet.[43]

⧝ ⧜

In 1998 18 people in Australia were seriously burned after trying on a new jumper with a lit cigarette in their mouth.[43]

⧝ ⧜

A massive 543 Australians were admitted to casualty in 1997 and 1998 after opening bottles of beer with their teeth or with their eye socket.[43]

⧝ ⧜

In 1998 5 people in Australia were injured in accidents involving out-of-control slot cars.[43]

⧝ ⧜

In 1997 8 Australians cracked their skull after passing out while throwing up into the toilet.[43]

⧝ ⧜

Brown fat burns well. It has been suggested that people who die of spontaneous human combustion may have had abnormal levels of brown fat in their bodies, making them more likely to literally "go up in flames".

⧝ ⧜

The odds of being killed by falling space debris are 1 in 5 billion.

⧝ ⧜

The odds of being killed by a dog are 1 in 700,000.

The odds of being killed by falling out of bed are 1 in 2 million.

If we had the same mortality rate now as in 1900, more than half the people in the world today would not be alive.

Famous last words: "It hurts" (French President Charles de Gaulle).

AFTERWORD

YET AGAIN, KEEP THOSE ODD BODY QUESTIONS COMING

I hope you have enjoyed reading *The Odd Body 3*. If so, and if you haven't done so already, you may wish to go back and read *The Odd Body* and *The Odd Body 2*. In some parts of the world, *The Essential Odd Body* combines these books. And if you still want more there are *The Odd Brain* and *The Odd Sex*. *The Odd Fear* is the next scheduled book in the "Odd" books series. And yes, there will be *The Odd Body 4* some time in the future.

If you have an Odd Body Question (OBQ) that has not been answered in earlier books, send it to me and I'll have a go at answering it, and might publish it in the next book. Send in more than 1 if you wish. I love your questions. Your name will be included if your question is chosen for the book. I'll even send you a free copy of the book — it's fun to give away books to all over the world. For the OBQ questioner it is a form of literary immortality!

Send your OBQs to:

Dr Stephen Juan
Faculty of Education and Social Work (A35)
University of Sydney
NSW 2006 AUSTRALIA

ACKNOWLEDGMENTS

Besides the author, many people are responsible for the production of *The Odd Body 3*. This is the section of a book where the author always tries to remember these people and where this author always succeeds in forgetting at least someone.

Thanks are owed to many authors for the wonderful research they've done that is mentioned in this book. The authors and their publishers have been listed either in the text or in the References.

Thanks to the staff at the University of Sydney libraries. They are always helpful.

Thanks to all the individuals who give me so much information either over the Internet, in print, over the airways when I'm on radio, by email, over the telephone or in person over coffee. Among these, a fount of information on so many domains of science is my good colleague Ian Stevens, of the Faculty of Education and Social Work at Sydney University. Over countless conversations, Ian has given me insights into many fields of science besides those in which I have been trained.

Thanks to the editors of newspapers around the world that run "The Odd Body" column each week, and the readers.

Thanks also to the book publishers around the world who have published foreign-language editions of my various books.

Thanks to radio and television station hosts and producers who continually invite me onto their programs to do what I do.

Many thanks are owed to my US literary agents, Muriel Nellis and Jane Roberts.

Many thanks, too, to the wonderful people at HarperCollins Publishers Australia, who always work so hard to make my books a success. Chief among these is Alison Urquhart, my publisher for several "Odd" books, almost since the series began. She is always the "proper" English lady, and I particularly enjoy our "proper" English teas at the Queen Victoria Building in Sydney. Thanks to Emma Kelso, my editor for *The Odd Body 3*, who offered suggestions, questions and an extra pair of eyes to help get this book as right as possible.

Others at HarperCollins work hard in editorial, design, typesetting, production, sales and publicity. I wish I knew all their names so I could thank them personally.

Research for this book, which occurred mostly on weekends, or early in the morning or late at night during the week, was undertaken while a staff member of the University of Sydney.

Of course, never forgotten are the readers who sent in OBQs that were used in this book: Heather Andrews, Paula Azur, T. Banfield, C. Barrie, Chris Bernard, Charles Bernie, F. Billington, Debbie Blair, Ian Bolger, Candace Bollinger, Jeremy Bono, Milos Brantke, Sophie Brown, Sarah Burgess, Will Burridge, Sandra Carlson, Paul Casey, Yvonne Chambers, Sonja Conners, Joe Cooper, Neil Cooper, Nicole Cooper, Kylie Cramer, David Crook, Jim Crossman, Stan Crowley, Jake Donner, S. Donohue, M. Downes, Rodney Downes, Connie Fitzgibbons, D. Fortier, Damien Fowler, Amy Francis, J. Gardner, Sam Gardner, Cathy Gibson,

Ronnie Gilbert, Matt Halliday, C. Hampton, Scotty Harper, Scott Harrison, B. Hawkins, Hans Heine, Dan Hemphill, Elizabeth Holland, John Hyland, Nathan James, Jason Janus, Anna Jeffes, Nicole Jeffes, Richard Jenkins, T. Jordan, Jenny Jordon, Ben Julian, Ben Kind, Jenny Knoll, E. Kocabas, Gillian Lacey, Jackie Lance, Martin Langford, Mark Lewis, Chantal Liebert, Jason Long, Richard McLain, Pat Maloney, James Marr, M. Marsh, Peter Martin, John Miller, Tom Millford, C. Monk, Nicole Murdoch, John Newton, Lucy Parker, Ralph Parslow, E. Perrins, Anne Perry, Laurie Perry, Sally Porter, James Raikar, Rory Rawlings, M. Reichburg, Glen Richman, Gary Robbins, S. Robinson, G. Rohrmann, Francis Salmeri, Barbara Sansome, Ken Scanlon, Hans Schmidt, Nathan Schuster, Ed Schwein, Kuranda Seyit, Prue Smith, R. Smith, Ian Smythe, D. Sonius, C. Stanhope, Simon Stevens, Jane Steward, Luke Stewart, Naomi Strossen, Kathy Swift, Bill Thayer, Chad Thomas, Peter Thomas, Mark Thompson, Rodney Thompson, Morgan Thring, Nathan Toms, Brad Townsend, Nicole Trudeau, S. Vance, Tommy Wallace, Kerry Waterman, Kathy Wellington, B. Williams, Dawn Williams, Lindy Williams and Hugh York.

Finally, thanks are given to my many colleagues and students at the University of Sydney.

Dr Stephen Juan

REFERENCES

CHAPTER 1: BEGINNINGS

1   S. Juan, "What is the probability of human life existing on other planets?", *National Post* (Toronto), 12 December 2005, "Body & Health", p. 1.

2   R. Leeuw, M. Cuttini, M. Nadai, I. Berbik, G. Hansen, A. Kucinskas, S. Lenoir, A. Levin, J. Persson, M. Rebagliato, M. Reid, M. Schroell and U. Vonderweid, "Treatment choices for extremely preterm infants: An international perspective", *Journal of Pediatrics*, 2000, vol. 137, no. 5, pp. 608–615.

3   L. Janus, *The Enduring Effects of the Prenatal Experience*, Jason Aronson, Northvale, New Jersey, 1997.

4   Dr Frans Veldman is a neonatalogist and is head of the International Centre for Research and Development of Haptonomy in Oms, Caret, France. He was the chief organiser of the first UNESCO conference on haptonomy in 1990.

5   F. Veldman, "Confirming affectivity, the dawn of human life", *International Journal of Prenatal and Perinatal Psychology and Medicine*, 1994, vol. 6, no. 1, pp. 11–26.

6   Personal communication, 15 July 1999.

7   Personal communication, 16 July 1999.

8   S. Juan, "Talking to the unborn", *Sun-Herald* (Sydney), 27 June 1999, "Tempo", p. 9.

9   S. Juan, "Hello in there", *New York Daily News*, 26 January 2005, "Body Work", p. 2.

10 M. Gustafson and P. Donahoe, "Male sex determination: Current concepts of male sexual differentiation", *Annual Review of Medicine*,

1994, vol. 45, pp. 505–524. Dr Gustafson and Dr Donohoe are from the Division of Pediatric Surgery at the Massachusetts General Hospital in Boston.

11 A. Montagu, *The Natural Superiority of Women* (3rd edition), Collier Books, New York, 1997, p. 80.

12 M. Gissler, M. Jarvelin, P. Louhiala and E. Hemminki, "Boys have more health problems in childhood than girls: Follow-up of the 1987 Finnish Birth Cohort", *Acta Paediatrica*, 1999, vol. 88, no. 3, pp. 310–340.

13 M. Robbins, "Nature, nurture, and core gender identity", *Journal of the American Psychoanalytic Association*, 1996, vol. 44 (supplement), pp. 93–117.

14 S. Juan, "Why it is so for males", *Sun-Herald* (Sydney), 25 July 1999, "Tempo", p. 16.

15 Personal communication, 30 August 2005.

16 T. Sadler, *Langman's Medical Embryology*, Lippincott Williams & Wilkins, Philadelphia, 2004.

17 S. Juan, "Is it possible for twins to have different fathers?", *National Post* (Toronto), 15 August 2005, "Body & Health", p. 2.

18 S. Juan, "Who was the world's tiniest baby?", *National Post* (Toronto), 15 August 2005, "Body & Health", p. 2.

19 H. Griengl, "Delusional pregnancy in a patient with primary sterility", *Journal of Psychosomatic Obstetrics and Gynaecology*, 2000, vol. 21, no. 1, pp. 57–59. Dr Griengl is a psychiatrist at the Vienna University Hospital in Austria.

20 S. Masoni, A. Maio, G. Trimarchi, C. de Punzio and P. Fioretti, "The couvade syndrome", *Journal of Psychosomatic Obstetrics and Gynaecology*, 1994, vol. 15, no. 3, pp. 125–131. Dr Masoni and colleagues are from the Department of Gynaecology and Obstetrics at the University of Pisa in Italy.

21 A. Storey, C. Walsh, R. Quinton and K. Wynne-Edwards, "Hormonal correlates of paternal responsiveness: New and expectant fathers", *Evolution and Human Behavior*, 2000, vol. 21, no. 2, pp. 79–95. Dr Storey and colleagues are from the Department of Psychology at Memorial University in Canada.

22 T. Field, M. Hernandez-Reif and J. Freedman, "Stimulation programs for preterm infants", *Social Policy Report* (Society for Research in Child Development), 2004, vol. 28, no. 1, pp. 3–19. Dr Field and her colleagues are from the Touch Research Institute of the School of Medicine at the University of Miami.

23 T. Field, "Massage improves disorders", *Brown University Child and Adolescent Behavior Letter*, December 1995, pp. 1–2.

24 S. Juan, "Rubbed the right way", *Sun-Herald* (Sydney), 2 July 2000, "Tempo", p. 14.

25 M. Zackrisson, "Brat pain", *New Scientist*, 6 July 1996, p. 65.

26 S. Juan, "When a baby cries . . .", *Sydney's Child*, February 2001, p. 38.

27 I. St James-Roberts and P. Menon-Johansson, "Predicting infant crying from fetal movement data: An exploratory study", *Early Human Development*, 1999, vol. 54, no. 1, pp. 55–62.

28 M. DeBellis, M. Keshavan, D. Clark, B. Casey, J. Giedd, A. Boring, K. Frustaci and N. Ryan, "Developmental traumatology, Part II: Brain development" (A.E. Bennett Research Award), *Biological Psychiatry*, 1999, vol. 45, no. 10, pp. 1271–1284. Dr DeBellis and colleagues are from the University of Pittsburgh.

29 M. Teicher, N. Dumont, Y. Ito, C. Vaituzis, J. Giedd and S. Andersen, "Childhood neglect is associated with reduced corpus callosum area", *Biological Psychiatry*, 2004, vol. 56, no. 2, pp. 80–85. Dr Teicher and colleagues are from Harvard University.

30 S. Juan, "Getting newborns to sleep peacefully", *National Post* (Toronto), 25 April 2005, "Body & Health", p. 1.

31 G. Anderson, E. Moore, J. Hepworth, and N. Bergman, "Early skin-to-skin contact for mothers and their healthy newborn infants", *Cochrane Database of Systematic Reviews*, 2003, vol. 2, CD003519. Dr Anderson and colleagues are from the Bolton School of Nursing at Case Western Reserve University in Cleveland, Ohio.

32 S. Juan, "Is skin-to-skin contact good for my newborn?", *National Post* (Toronto), 25 April 2005, "Body & Health", p. 1.

33 A. Taddio, V. Shah, C. Gilbert-MacLeod and J. Katz, "Conditioning and hyperalgesia in newborns exposed to repeated heel lances", *Journal of the American Medical Association*, 2002, vol. 288, no. 7, pp. 857–861. Dr Taddio and her colleagues are from the Hospital for Sick Children in Toronto, Canada.

34 S. Juan, "Can newborn babies anticipate pain?", *National Post* (Toronto), 24 October 2005, "Body & Health", p. 2.

35 L. Salk, L. Lipsitt, W. Sturner, B. Reilly and R. Levat, "Relationship of maternal and perinatal conditions to eventual adolescent suicide", *The Lancet*, 1985, vol. 1, pp. 624–627. Dr Salk and colleagues are from Cornell University Medical College, New York, and Brown University, Rhode Island.

36 B. Jacobson, K. Nyberg, L. Grönbladh, G. Eklund, M. Bygdeman and U. Rydberg, "Opiate addiction in adult offspring through possible imprinting after obstetric treatment", *British Medical Journal*, 1990, vol. 301, pp. 1067–1070. Dr Jacobson and colleagues are from the Karolinska Insititute in Stockholm.

37 A. Raine, P. Brennan and S. Mednick, "Interaction between birth complications and early maternal rejection in predisposing individuals to adult violence: Specificity to serious, early-onset violence", *American Journal of Psychology*, 1997, vol. 154, no. 9, pp. 1265–1271. Dr Raine and colleagues are from the department of Psychology at the University of Southern California.

38 M. Odent, *The Caesarean*, Free Associations Books, London, 2004. Dr Odent, a world-famous French obstetrician, is the founder of the Primal Health Research Centre in London.

39 S. Juan, "Mind over mammaries", *National Post* (Toronto), 19 March 2005, "Body & Health", p. 1.

40 S. Juan, "Why do we have two of so many body parts?", *National Post* (Toronto), 25 July 2005, "Body & Health", p. 2.

41 American Dental Association website, 12 August 2004.

42 C. Sinnatamby (ed.), *Last's Anatomy: Regional and Applied* (10th edition), Churchill Livingstone, Edinburgh, 1999.

43 S. Juan, "Is there any part of the body we don't need?", *New York Daily News*, 27 October 2004, "Body Work", pp. 1–2.

44 S. Juan, "Who are you calling obsolete the appendix may ask", *National Post* (Toronto), 4 November 2004, "Body & Health", p. 1.

45 S. Olshansky, B. Carnes and R. Butler, "If humans were built to last", *Scientific American*, March 2001, pp. 50–55. Dr Olshansky and colleagues are from the School of Public Health at the University of Illinois in Chicago.

46 S. Juan, "How was the Neanderthal man different from us?", *National Post* (Toronto), 8 August 2005, "Body & Health", p. 1.

47 Dr Hart is from the Department of Anthropology at the University of Missouri in St Louis and Dr Sussman is from the Department of Anthropology at Washington University in St Louis.

48 D. Hart and R. Sussman, *Man the Hunted: Primates, Predators and Human Evolution*, Westview Press, Boulder, Colorado, 2005.

49 Dr Ray Kurzweil is a physicist, engineer, author and an independent scientist formerly at the Massachusetts Institute of Technology (MIT).

50 R. Kurzweil and T. Grossman, *Fantastic Voyage: Live Long Enough to Live Forever*, Rodale Press, New York, 2004.

51 R. Kurzweil, *The Age of Spiritual Machines: When Computers Exceed Human Intelligence*, Viking Books, New York, 1999.

CHAPTER 2: THE HEAD

1 A. Phelps, "Head trauma", *New Scientist*, 16 November 2002, p. 65.

2   Dr V. Riccieri and Dr G. Valesini are from the Rheumatology Centre of the Department of Clinical Therapy and Applied Medicine at the University of La Sapienza in Rome.

3   V. Riccieri and G. Valesini, "Treatment of Wegener's granulomatosis", *Reumatismo*, 2004, vol. 56, no. 2, pp. 69–76.

4   Dr A. Todorov, "A baby face seems bad in politics", *Newsday*, 10 June 2005, p. 14.

5   S. Juan, "Do facial features affect personality?", *National Post* (Toronto), 22 August 2005, "Body & Health", p. 1.

6   Dr Mark Hans is from the School of Dentistry at Case Western Reserve University in Cleveland, Ohio.

7   M. Hans, S. Nelson, N. Pracharktam, S. Baek, K. Strohl and S. Redline, "Subgrouping persons with snoring and/or apnoea by using anthropometric and cephalometric measures", *Sleep and Breathing*, 2001, vol. 5, no. 2, pp. 79–91.

8   S. Juan, "Are you a snorehead?" *New York Daily News*, 1 September 2004, "Body Work", p. 1.

9   Dr M. Valenca and colleagues are from the Division of Neurology and Neurosurgery in the Department of Neuropsychiatry at the Federal University of Pernambuco in Recife, Brazil.

10 M. Valenca, L. Valenca, C. Bordini, W. da Silva, J. Leite, J. Antunes-Rodrigues and J. Speciali, "Cerebral vasospasm and headache during sexual intercourse and masturbatory orgasms", *Headache*, 2004, vol. 44, no. 3, pp. 244–248.

11 Dr John Ostergaard is one of the pioneers of benign coital headache syndrome research.

12 J. Ostergaard, "Natural course of benign coital headache", *British Medical Journal*, 1992, vol 305, p. 1129.

13 S. Juan, "Can sex cause a headache?", *New York Daily News*, 2 March 2005, "Body Work", p. 1.

14 S. Juan, "Can sex cause a headache?", *National Post* (Toronto), 14 March 2005, "Body & Health", p. 1.

15 S. Juan, *The Odd Body: Mysteries of our Weird and Wonderful Bodies Explained*, HarperCollins, Sydney, 1995, p. 43.

16 S. Juan, "Plate tectonics", *New York Daily News*, 4 October 2005, "Body Work", p. 1.

17 Dr John Meara is Director of the Department of Plastic and Maxillofacial Surgery at the Royal Children's Hospital in Melbourne.

18 J. Meara, "Deformational plagiocephaly", *Community Paediatric Review*, 2004, vol. 13, no. 2, pp. 1–3.

19 S. Juan, "Is it possible to be born with two heads and survive?", *National Post* (Toronto), 8 August 2005, "Body & Health", pp. 1–2.

20 Dr M. Ernberg and colleagues are from the Karolinska Institutet in Stockholm, Sweden.

21 M. Ernberg, B. Hedenberg-Magnusson, P. Alstergren, T. Lundeberg and S. Kopp, "Pain, allodynia, and serum serotonin level in orofacial pain of muscular origin", *Journal of Orofacial Pain*, 1999, vol. 13, no. 1, pp. 56–62.

22 Dr G.B. Wexler and Dr M.W. McKinney are from the Ottawa Civic Hospital in Ottawa, Ontario, Canada.

23 G. Wexler and M. McKinney, "Temporomandibular treatment outcomes with five diagnostic categories", *Cranio*, 1999, vol. 17, no. 1, pp. 30–37.

24 "TMD spells jaw pain", *UC Wellness Letter*, March 1994, pp. 6–7.

25 S. Juan, "No time to talk", *Sun-Herald* (Sydney), 7 November 1999, "Tempo", p. 15.

26 S. Blakeslee, "The mystery of Mona Lisa's smile linked to flickering eyes", *New York Times*, 27 November 2000, p. 81.

27 K. Svitil, "Mona Lisa smile", *Discover*, June 2003, p. 15.

28 S. Juan, "Why does the smile of the Mona Lisa seem to appear and then disappear?", *National Post* (Toronto), 11 October 2005, "Body & Health", p. 1.

29 T. Sterling, "Scientists figure out why Mona Lisa smiles", *Associated Press*, 15 December 2005.

30 P. Ekman, *Emotions Revealed: Recognising Faces and Feelings to Improve Communication and Emotional Life*, Henry Holt, New York, 2004.

31 S. Juan, "Is it possible to 'read' a person's face?", *National Post* (Toronto), 28 November 2005, "Body & Health", p. 1.

32 S. Juan, "Why do they call plastic surgery 'plastic'?", *National Post* (Toronto), 19 December 2005, "Body & Health", pp. 1–2.

33 S. Juan, "Synthetic people", *Sun-Herald* (Sydney), 2 May 1999, "Tempo", p. 16.

34 S. Juan, "What is the Pinocchio effect?", *New York Daily News*, 2 August 2005, "Body Work", pp. 1–2.

35 S. Juan, "What is the Pinocchio effect?", *National Post* (Toronto), 22 August 2005, "Body & Health", pp. 1–2.

CHAPTER 3: THE EYES

1 D. Apple and M. Rabb, *Ocular Pathology* (5th edition), Mosby, St Louis, 1998.

2 Dr David J. Apple is from the Moran Center of the University of Utah Health Sciences Center in Salt Lake City.

3   A. Coulombre, "The role of intraocular pressure in the development of the chick eye", *Journal of Experimental Zoology*, 1956, vol. 133, pp. 211–225.

4   R. Goss, *The Physiology of Growth*, Academic Press, New York, 1978, pp. 191–199.

5   S. Juan, "Do my eyeballs remain the same size from birth?", *National Post* (Toronto), 22 August 2005, "Body & Health", p. 2.

6   R. Moses and W. Hart (eds), *Adler's Physiology of the Eye* (10th edition), Mosby, St Louis, 2003.

7   S. Juan, "The eyes have it", *New York Daily News*, 8 September 2004, "Body Work", p. 2.

8   Dr Tom Stickel is from the School of Optometry at Indiana University in Bloomington.

9   S. Schwartz, *Visual Perception: A Clinical Orientation*, Appleton & Lange, Norwalk, Connecticut, 1994.

10 S. Juan, "How long does it take to damage your eyes when you stare at the sun?", *National Post* (Toronto), 11 October 2005, "Body & Health", pp. 1–2.

11 Dr John Morenski is from the Division of Neurosurgery at the University of Missouri in Columbia.

12 S. Juan, "Why do I sometimes visualize a glowing green or yellow 'eye' after rubbing my eyes?", *National Post* (Toronto), 16 January 2006, "Body & Health", pp. 1–2.

13 N. Lessard, M. Pare, F. Lepore and M. Lassonde, "Early-blind human subjects localize sound souces better than sighted subjects", *Nature*, 1998, vol. 395, pp. 278–280.

14 N. Lessard, F. Lepore, J. Villemagne and M. Lassonde, "Sound localization in callosal agenesis and early callosotomy subjects: Brain reorganization and compensatory strategies", *Brain*, 2002, vol. 125, pp. 1039–1053.

15 K. Pallarito, "Early blindness sharpens sense of sound", *Daily Health and Medical News*, 18 August 2004.

16 K. Wuensch, R. Chia, W. Castellow, C. Chuang and B. Cheng, "Effects of physical attractiveness, sex, and type of crime on mock juror decisions: A replication with Chinese students", *Journal of Cross-Cultural Psychology*, 1993, vol. 24, pp. 414–427.

17 K. Wuensch and C. Moore, "Effects of physical attractiveness on evaluations of a male employee's allegations of sexual harassment by his female employer", *Journal of Social Psychology*, 2004, vol. 144, pp. 207–211.

18 Dr D. Abwender and Dr K. Hough are from the Department of Psychology at the State University of New York at Brockport.

19 D. Abwender and K. Hough, "Interactive effects of characteristics of defendant and mock juror on US participants' judgment and sentencing recommendations", *Journal of Social Psychology*, 2001, vol. 141, pp. 603–615.

20 S. Juan, "Unique eye-dentity", *New York Daily News*, 5 April 2005, "Body Work", p. 1.

21 S. Juan, "Iris, retina hold keys to eye-dentification", *National Post* (Toronto), 9 May 2005, "Body & Health", p. 1.

22 S. Juan, "How does peripheral vision work?", *National Post* (Toronto), 24 October 2005, "Body & Health", p. 2.

23 Dr Tom Wilson is a pathologist from the Division of Molecular Oncology at the Washington University School of Medicine in St Louis, Missouri.

24 J. Masson and S. McCarthy, *When Elephants Weep: The Emotional Lives of Animals*, Wheeler, Rockland, Maryland, 1995.

25 S. Juan, "Animals 'cry' when they are separated", *National Post* (Toronto), 14 November 2005, "Body & Health", pp. 1–2.

26 S. Juan, "Are humans the only animals that cry?", *New York Daily News*, 23 November 2005, "Body Work", pp. 1–2.

27 R. von der Heydt, "Image parsing mechanisms of the visual cortex", in J. Werner and L. Calupa (eds), *The Visual Neurosciences*, MIT Press, Cambridge, Massachusetts, 2003, pp. 1139–1150.

28 R. Sheldrake, *The Sense of Being Stared*, Three Rivers Press, New York, 2003.

29 G. Rosenthal, B. Soper, E. Folse and G. Whipple, "Ability to detect covert observations", *Perceptual & Motor Skills*, 1997, vol. 85, p. 75.

30 Dr C.M. Cook and Dr M.A. Persinger are from the Department of Neuroscience at Laurentian University in Ontario, Canada.

31 C. Cook and M. Persinger, "Experimental induction of the 'sensed presence' in normal subjects and an exceptional subject", *Perceptual & Motor Skills*, 1997, vol. 85, p. 683.

32 S. Juan, "Do supermarkets hypnotize you to shop?", *National Post* (Toronto), 12 December 2005, "Body & Health", pp. 1–2.

33 S. Juan, "The ghost in the machine", *New York Daily News*, 28 September 2004, "Body Work", p. 2.

34 S. Juan, "Synthetic people", *Sun-Herald* (Sydney), 2 May 1999, "Tempo", p. 16.

CHAPTER 4: THE NOSE

1 S. Juan, "The psychology of smell", *Sun-Herald* (Sydney), 4 April 1999, "Tempo", p. 13.

2   Personal communication, 12 March 1999.

3   S. Juan, "Can the smell in a supermarket affect consumer spending?",
    National Post (Toronto), 11 October 2005, "Body & Health", p. 2.

4   S. Juan, "Cinnamon buns take the cake with smell", *National Post*
    (Toronto), 17 October 2005, "Body & Health", pp. 1–2.

5   Dr R. James Swanson is from the Department of Biological Sciences,
    Obstetrics and Gynecology at Old Dominion University in Norfolk,
    Virginia.

6   S. Juan, "Cat hair has no effect on nasal hair", *National Post*
    (Toronto), 9 January 2006, "Body & Health", p. 1.

7   Carlos Padilla is from the University Clinic at the University of Texas
    Health Science Center in Austin.

8   S. Juan, "Can the cold cause a cold?", *National Post* (Toronto),
    14 March 2005, "Body & Health", pp. 1–2.

9   S. Juan, "What is the Pinocchio effect?", *New York Daily News*,
    2 August 2005, "Body Work", pp. 1–2.

10 Dr Alan Hirsch is from the Smell and Taste Treatment and Research
    Foundation in Chicago.

11 S. Juan, "What is the Pinocchio effect?", *National Post* (Toronto),
    22 August 2005, "Body & Health", pp. 1–2.

12 C. Wessner, "FYI", *Popular Science*, January 2000, p. 83.

13 S. Mitchell, "Food idiosyncrasies: Beetroot and asparagus", *Drug
    Metabolism and Disposition*, 2001, vol. 29, no. 4 (pt 2), pp. 539–543.
    Dr Mitchell is from the Division of Biomedical Sciences in the
    Faculty of Medicine at the Imperial College in London.

14 A. Matluk, "Scent of a man", *New Scientist*, 10 February 2001,
    pp. 34–38.

15 S. Juan, "How important is histocompatibility (compatible immune
    systems) in human mating?", *New York Daily News*, 6 December
    2005, "Body Work", pp. 1–2.

16 S. Juan, "Sniffing out an ideal mate", *National Post* (Toronto),
    2 January 2005, "Body & Health", pp. 1–2.

17 S. Juan, "Why do they call it 'hay fever'?", *National Post* (Toronto),
    31 October 2005, "Body & Health", p. 2.

CHAPTER 5: THE EARS

1   S. Juan, "Why can't I wiggle my ears?", *National Post* (Toronto),
    14 March 2005, "Body & Health", p. 1.

2   S. Juan, "Sifting through the din", *National Post* (Toronto),
    19 September 2005, "Body & Health", p. 1.

3   Dr Lloyd Tripp is a psychologist from the Department of Human
    Factors Experimental Psychology at the University of Cincinnati.

4 S. Juan, "Staying vertical", *New York Daily News*, 2 February 2005, "Body Work", p. 1.

5 S. Juan, "Why can't humans hear radio waves?", *National Post* (Toronto), 11 October 2005, "Body & Health", p. 2.

6 Dr Andrea Zardetto-Smith is from the Department of Physical and Occupational Therapy and Biomedical Sciences at Creighton University in Omaha, Nebraska.

7 M. Bear, B. Connors and M. Paradiso, *Neuroscience: Exploring the Brain*, William and Wilkins, Baltimore, Maryland, 1996.

8 P. Hofman, J. Van Riswick and A. Van Opstal, "Relearning sound localization with new ears", *Nature Neuroscience*, 1998, vol. 1, no. 5, pp. 417–421.

9 R. Salvi, "An inescapable buzz", *Discover*, October 1995, p. 28.

10 S. Juan, "Synthetic people", *Sun-Herald* (Sydney), 2 May 1999, "Tempo", p. 16.

CHAPTER 6: THE MOUTH

1 Dr Richard Wiseman is a psychologist at the Laugh Lab of the University of Hertfordshire in the UK.

2 S. Juan, "Only when I laugh", *Sun-Herald* (Sydney), 3 October 1999, "Tempo", p. 14.

3 Dr Charles B. Simpson is from the Department of Otolaryngology at the University of Texas Health Science Center at San Antonio.

4 S. Juan, "Voices of the people", *New York Daily News*, 21 July 2004, "Body Work", p. 2.

5 Dr M. L. Lalakea and Dr A. H. Messner are from the Division of Surgery, Otolaryngology, Head and Neck Surgery at Stanford University.

6 C. Kucik, G. Martin and B. Sortor, "Common intestinal parasites", *American Family Physician*, 2004, vol. 69, no. 5, pp. 1161–1168.

7 Dr T.J. Wilkinson is a chemist and forensic scientist from the Department of General Chemistry and Forensic Science at the Lawrence Berkeley National Laboratory in Berkeley, California.

8 S. Juan, "Do lips have prints?", *New York Daily News*, 16 February 2005, "Body Work", p. 1.

9 S. Juan, "Do lips have prints?", *National Post* (Toronto), 11 March 2005, "Body & Health", p. 1.

10 Personal communication, 13 January 2005, with Dr David Barnett and his teamate at the Stranmillis University College in Belfast, Northern Ireland.

11 C. Clayton, *Pyjamarama: Sleepover Handbook*, Bloomsbury, London, 1996.

12 S. Juan, "The things people do in the night", *National Post* (Toronto), 11 July 2005, "Body & Health", pp. 1–2.

13 M. Hans, S. Nelson, N. Pracharktam, S. Baek, K. Strohl and S. Redline, "Subgrouping persons with snoring and/or apnea by using anthropometric and cephalometric measures", *Sleep and Breathing*, 2001, vol. 5, no. 2, pp. 79–91.

14 S. Juan, "What is snoring and why do people who snore not hear their own snoring and wake up?", *National Post* (Toronto), 16 January 2006, "Body & Health", p. 1.

15 S. Juan, "It's all a matter of taste", *New York Daily News*, 19 April 2005, "Body Work", p. 1.

16 Dr Vidya Bhalodia is from the Department of Neuroscience at Washington University in St Louis, Missouri.

17 P. Waddell, "Burning question", *New Scientist*, 16 December 1995, p. 65.

18 Dr Theodore Levin and Dr Michael Edgerton are from the Vocal Functions Laboratory at the University of Wisconsin in Madison.

19 M. Lalakea and A. Messner, "Ankyloglossia: Does it matter?", *Pediatric Clinics of North America*, 2003, vol. 50, no. 2, pp. 381–397.

20 S. Juan, "It's not just a matter of taste", *New York Daily News*, 25 August 2004, "Body Work", p. 2.

21 A. Kupietzky and E. Botzer, "Ankyloglossia in the infant and young child: Clinical suggestions for diagnosis and management", *Pediatric Dentistry*, 2005, vol. 27, pp. 40–46. Dr Kupietzky and Dr Botzer are from the Hebrew University-Hadassah School of Dental Medicine in Jerusalem.

22 R. Fiorotti, M. Bertolini, J. Nicola and E. Nicola, "Early lingual frenectomy assisted by CO_2 laser helps prevention and treatment of functional alterations caused by ankyloglossia", *International Journal of Orofacial Myology*, 2004, vol. 30, pp. 64–71. Dr Fiorotti, Dr Bertolini, Dr Nicola and Dr Nicola are from the School of Medical Sciences at the State University of Campinas in Brazil.

23 S. Juan, "A special lubricant", *New York Daily News*, 25 July 2005, "Body Work", pp. 1–2.

24 S. Juan, "Bad breath wouldn't be an issue if we drooled like babies", *National Post* (Toronto), 7 October 2004, "Body & Health", p. 1.

25 S. Juan, "An infant phenomenon", *New York Daily News*, 13 October 2004, "Body Work", p. 2.

26 Dr Steven Pinker is Professor of Cognitive Sciences at the Massachusetts Institute of Technology (MIT) and author of *The Language Instinct* (HarperCollins, New York, 1994).

27 Dr O. Amir and Dr L. Kishon-Rabin are from the Department of

Communication Disorders at the Chaim Sheba Medical Center, Tel-Aviv University.

28 O. Amir and L. Kishon-Rabin, "Association between birth control pills and voice quality", *Laryngoscope*, 2004, vol. 114, no. 6, pp. 1021–1026.

29 Dr M.M. Gorham-Rowan and colleagues are from the Communications Disorders Program at Georgia State University in Atlanta.

30 M. Gorham-Rowan, A. Langford, K. Corrigan and B. Snyder, "Vocal pitch levels during connected speech associated with oral contraceptive use", *Journal of Obstetrics and Gynaecology*, 2004, vol. 24, no. 3, pp. 284–286.

31 S. Juan, "Synthetic people", *Sun-Herald* (Sydney), 2 May 1999, "Tempo", p. 16.

CHAPTER 7: THE SKIN

1 S. Juan, "Navel gazing", *Sun-Herald* (Sydney), 10 September 2000, "Tempo", p. 14.

2 S. Juan, "A history of navel gazing", *New York Daily News*, 7 July 2004, "Body Work", p. 2.

3 Personal communication, 12 June 2000.

4 Dr Richard Polin is a Professor of Paediatrics at the Columbia University Medical School in New York and Director of the Neonatal Unit at the Children's Hospital of the New York Presbyterian Hospital.

5 C. Ray, "Navel perspective", *New York Times*, 15 April 2003, p. D2.

6 S. Juan, "Left red-faced", *New York Daily News*, 28 July 2004, "Body Work", p. 2.

7 L. Carroll and R. Anderson, "Body piercing, tattooing, self-esteem, and body investment in adolescent girls", *Adolescence*, 2002, vol. 37, no. 147, pp. 627–637. Dr Carroll and Dr Anderson are from the University of North Florida in Jacksonville.

8 S. Bridgeman-Shah, "The medical and surgical therapy of pseudofolliculitis barbae", *Dermatologic Therapy*, 2004, vol. 17, no. 2, pp. 158–163. Dr Bridgeman-Shah is from the Department of Dermatology at Howard University in Washington, D.C.

9 J. Garcia-Zuazaga, "Pseudofolliculitis barbae: Review and update of new treatment modalities", *Military Medicine*, 2003, vol. 168, no. 7, pp. 561–564.

10 V. Jain, "Markism", *New Scientist*, 22 February 2003, p. 65.

11 M. Goldwyn, *How a Fly Walks Upside Down ... And Other Curious Facts*, Wings Books, New York, 1995, p. 234.

12 S. Juan, "What are freckles and what causes them?", *New York Daily News*, 3 May 2005, "Body Work", p. 1.

13 I. Johnson, *Why Can't You Tickle Yourself and Other Bodily Curiosities*, Warner Books, New York, 1993, p. 11.

14 S. Juan, "Bruise is a sign of small blood vessel leakage", *National Post* (Toronto), 15 August 2005, "Body & Health", p. 1.

15 S. Juan, "What is a bruise?", *New York Daily News*, 24 August 2005, "Body Work", p. 1.

16 M. Goldwyn, *How a Fly Walks Upside Down ... And Other Curious Facts*, Wings Books, New York, 1995, p. 173.

17 A. Xenakis, *Why Doesn't My Funny Bone Make Me Laugh?*, Villard Books, New York, 1993, p. 155.

18 C. Ray, "Suntans", *New York Times*, 11 May 1999, p. D2.

19 S. Juan, "A special lubricant", *National Post* (Toronto), 25 July 2005, "Body & Health", pp. 1–2.

20 S. Juan, "The skinny on skin", *New York Daily News*, 23 February 2005, "Body Work", p. 1.

21 Dr Nina Jablonski is Irvine Chair and Curator of Anthropology at the California Academy of Sciences in San Francisco.

22 G. Rich, "Healing hands", *Psychology Today*, March/April 1999, p. 23.

23 E. Larsen, "Do touch: The benefits of skin on skin go deeper than feeling good", *Utne Reader*, March–April 1998, pp. 78–81.

24 D. Eller, "Rubbed the right way", *American Health*, January/February 1996, pp. 74–77.

25 S. Juan, "White fright", *New York Daily News*, 13 April 2005, "Body Work", p. 2.

26 S. Juan, "Why do we 'turn white with fright'?", *National Post* (Toronto), 19 March 2005, "Body & Health", pp. 1–2.

27 M. Stucker, A. Struk, P. Altmeyer, M. Herde, H. Baumgartl and D. Lubbers, "The cutaneous uptake of atmospheric oxygen contributes significantly to the oxygen supply of human dermis and epidermis", *Journal of Physiology*, 2004, vol. 538 (pt 3), pp. 985–994. Dr Markus Stucker and his colleagues are from Ruhr-University Bochum in Bochum, Germany.

28 D. Mooney and A. Mikos, "Growing new organs", *Scientific American*, April 1999, pp. 60–65. Dr David Mooney is an Associate Professor of Biology at the University of Michigan in Ann Arbor and Dr Antonios Mikos is an Associate Professor of Bioengineering at Rice University in Houston, Texas.

29 S. Juan, "Synthetic people", *Sun-Herald* (Sydney), 2 May 1999, p. 16.

CHAPTER 8: THE HAIR & NAILS

1 D. Morris, *The Naked Ape: A Zoologist's Study of the Human Animal*, McGraw-Hill, New York, 1967.

2 S. Juan, "Why does hair on my head grow longer than hair anywhere else on my body?", *New York Daily News*, 11 August 2004, "Body Work", p. 2.

3 D. Tran and R. Sinclair, "Understanding and managing common baldness", *Australian Family Physician*, 1999, vol. 28, no. 3, pp. 248–253.

4 D. Nelson, "Aaaaaargh", *New Scientist*, 11 April 1998, p. 64.

5 S. Juan, "White fright", *New York Daily News*, 13 April 2005, "Body Work", p. 2.

6 S. Juan, "Can hair turn white overnight as the result of shock?", *National Post* (Toronto), 9 May 2005, "Body & Health", pp. 1–2.

7 S. Juan, "Why do we have eyelashes?", *New York Daily News*, 21 June 2005, "Body Work", pp. 1–2.

8 C. Darwin, *The Descent of Man*, J. Murray, London, 1871.

9 E. Morgan, *The Aquatic Ape*, Stein and Day, New York, 1982.

10 M. Harris, *Our Kind: Who We Are, Where We Came From & Where We Are Going*, HarperCollins, New York, 1989.

11 S. Juan, "Why we are all nearly bald", *Sun-Herald* (Sydney), 28 May 2000, "Tempo", p. 12.

12 S. Juan, "Why are humans nearly bald? (Part 1)", *National Post* (Toronto), 12 September 2005, "Body & Health", pp. 1–2.

13 S. Juan, "Why are humans nearly bald? (Part 2)", *National Post* (Toronto), 19 September 2005, "Body & Health", pp. 1–2.

14 D. Tran and R. Sinclair, "Understanding and managing common baldness", *Australian Family Physician*, 1999, vol. 28, no. 3, pp. 248–253.

15 P. Wells, T. Willmoth and R. Russell, "Does fortune favour the bald? Psychological correlates of hair loss in males", *British Journal of Psychology*, 1995, vol. 86, pp. 337–344.

16 C. Gosselin, "Hair loss, personality and attitudes", *Personality and Individual Differences*, 1984, vol. 5, pp. 365–369.

17 L. Sigelman, E. Dawson, M. Nitz and M. Whicker, "Hair loss and electability: The bald truth", *Journal of Nonverbal Behavior*, 1990, vol. 14, pp. 269–283.

18 S. Juan, "The bald and the beautiful", *Sun-Herald* (Sydney), 23 May 1999, "Tempo", p. 17.

19 Dr Angela Christiano is from the Columbia-Presbyterian Medical Center in New York.

20 J. Knight, "Keep your hair on", *New Scientist*, 7 February 1998, p. 15.

21 C. Ray, "Genes and baldness", *New York Times*, 14 September 2004, p. D2.
22 Dr L. Bry is from the Department of Pathology at the Brigham and Women's Hospital and Harvard Medical School in Cambridge, Massachusetts.
23 S. Juan, "Our handy digits", *National Post* (Toronto), 11 April 2005, "Body & Health", p. 1.

CHAPTER 9: THE SKELETON, BONES & TEETH

1 Dr Diane Kelly is from the Department of Biomedical Sciences at Cornell University in New York.
2 Dr Paul Odgren is from the Department of Cell Biology at the University of Massachusetts Medical School in Worcester.
3 S. Juan, "Snap, crackle, pop!", *New York Daily News*, 30 March 2005, "Body Work", p. 2.
4 Personal communication, 6 March 2005. Dr Peter Bonafede is Medical Director of the Providence Arthritis Center at Providence Portland Medical Center in Rhode Island.
5 Dr P. Chan is from the Department of Orthopedics at the University of Pennsylvania Medical Center in Philadelphia.
6 P. Chan, D. Steinberg and D. Bozentka, "Consequences of knuckle cracking: A report of two acute injuries", *American Journal of Orthopedics*, 1999, vol. 28, no. 2, pp. 113–114.
7 A. Guyton and J. Hall, *Pocket Companion of Textbook of Medical Physiology* (10th edition), W.B. Saunders, St Louis, 2001, pp. 851–854.
8 Dr John Morenski is from the Division of Neurosurgery at the University of Missouri in Columbia.
9 Dr S. Managoli, Dr P. Chaturvedi, Dr K. Vilhekar and Dr J. Ivenger are from the Department of Pediatrics, Mahatma Gandhi Institute of Medical Sciences, Sevagram, Wardha, Maharashtra, India.
10 S. Managoli, P. Chaturvedi, K. Vilhekar and J. Ivenger, "Mermaid syndrome with amniotic band disruption", *Indian Journal of Pediatrics*, 2003, vol. 70, no. 1, pp. 105–107.
11 K. Taori, K. Mitra, N. Ghonge, R. Gandhi, T. Mammen and J. Sahu, "Sirenomelia sequence: Report of three cases", *Indian Journal of Radiological Imaging*, 2002, vol. 12, no. 3, pp. 399–401.
12 Dr Christine Rodda is Head of the Department of Paediatric Endocrinology and Diabetes at the Monash Medical Centre in Clayton, Victoria.
13 C. Rodda, "Rickets in the 21st century", *Community Paediatric Review*, 2005, vol. 14, no. 2, pp. 1–4.

14 Dr Paul Odgren is from the Department of Cell Biology at the University of Massachusetts Medical School in Worcester.

15 Dr Michael Dougherty is from the Biological Sciences Curriculum Study Project in Colorado Springs, Colorado.

16 The late Dr S. Feldman worked in the Department of Psychology at Cornell University in Ithaca, New York.

17 S. Feldman, "Phantom limb", *American Journal of Psychology*, 1940, vol. 53, pp. 590–592.

18 R. Melzack, "Phantom limbs", *Scientific American*, January 1992, pp. 90–96. Dr Melzack is from the Department of Psychology at McGill University in Montreal, Canada.

19 V. Ramachandran and S. Blakeslee, *Phantoms of the Brain: Probing the Mysteries of the Human Mind*, William Morrow, New York, 1998. Dr V.S. Ramachandran is from the Department of Psychology at the University of California in San Diego.

20 S. Juan, "Out on a limb", *Sun-Herald* (Sydney), 18 July 1999, "Tempo", p. 15.

21 S. Juan, "May the circle be unbroken", *New York Daily News*, 17 May 2005, "Body Work", pp. 1–2.

22 S. Juan, "Why we go around in circles", *National Post* (Toronto), 6 June 2005, "Body & Health", pp. 1–2.

23 Personal communication, 14 March 2004. Dr Robert Houska is from the Department of Natural Sciences at Fullerton College in Fullerton, California.

24 C. Dugmore and W. Rock, "The prevalence of tooth erosion in 12-year-old-children", *British Dental Journal*, 2004, vol. 196, no. 5, pp. 279–282.

25 S. Juan, "Choppers down", *New York Daily News*, 19 July 2005, "Body Work", pp. 1–2.

26 T. Ooshima, Y. Osaka, H. Sasaki, K. Osawa, H. Yasuda and M. Matsumoto, "Cariostatic activity of cacao mass extract", *Archives of Oral Biology*, 2000, vol. 45, no. 9, pp. 805–808.

27 S. Goforth, "Test tube teeth", *The Why? Files of the University of Wisconsin*, 19 February 2004. Dr Paul Sharpe is from the Dental Institute at King's College in London.

28 S. Juan, "Decayed, lost and re-grown", *National Post* (Toronto), 18 July 2005, "Body & Health", pp. 1–2.

29 D. Mooney and A. Mikos, "Growing new organs", *Scientific American*, April 1999, pp. 60–65.

30 L. Iannucci, "Brave new bodies: The future is here", *American Health*, December 1995, pp. 68–69.

31 S. Juan, "Synthetic people", *Sun-Herald* (Sydney), 2 May 1999, "Tempo", p. 16.

32 Dr Robert Langer is from the Massachusetts Institute of Technology in Cambridge, Massachusetts.

33 S. Juan, "Our handy digits", *National Post* (Toronto), 11 April 2005, "Body & Health", p. 1.

CHAPTER 10: THE HEART, BLOOD & LUNGS

1 M. Fishman and K. Chien, "Fashioning the vertebrate heart: Earliest embryonic decisions", *Development*, 1997, vol. 124, pp. 2099–2117.

2 T. Mohun and D. Sparrow, "Early steps in vertebrate cardiogenesis", *Current Opinion in Genetics & Development*, 1997, vol. 7, pp. 628–633.

3 S. Juan, "What causes the human heart to first start beating?", *National Post* (Toronto), 15 August 2005, "Body & Health", pp. 1–2.

4 S. Juan, "Is it true that the younger you are the faster your heart beats?", *National Post* (Toronto), 12 September 2005, "Body & Health", p. 2.

5 Dr Terry Hebert is a heart specialist at the Department of Biochemistry at the Montreal Heart Institute of the University of Montreal.

6 S. Juan, "Mighty muscle launch", *New York Daily News*, 6 October 2004, "Body Work", p. 2.

7 Dr Peter Scheufele is from the Uniformed Services University of the Health Sciences in Bethesda, Maryland.

8 P. Scheufele, "Effects of progressive relaxation and classical music on measurements of attention, relaxation, and stress responses", *Journal of Behavioral Medicine*, 2000, vol. 23, no. 2, pp. 207–228.

9 Dr Eric Tardif is from the Institut de Physiologie at the Université de Lausanne in Switzerland.

10 M. Iwanaga and M. Tsukamoto, "Effects of excitative and sedative music on subjective and physiological relaxation", *Perceptual & Motor Skills*, 1997, vol. 85, no. 1, pp. 287–296.

11 *Pravda*, "A hearty afterlife", reported in *Fortean Times*, June 2005, p. 10.

12 Dr G. Monreal is from the Department of Cardiothoracic Surgery at the School of Medicine at Ohio State University in Columbus.

13 S. Gould, *The Panda's Thumb*, W.W. Norton, New York, 1980.

14 S. Juan, "Your life in heartbeats", *New York Daily News*, 1 June 2005, "Body Work", pp. 1–2.

15 S. Juan, "The ins and outs of our remarkable lungs", *National Post* (Toronto), 13 June 2005, "Body & Health", p. 2.

16 A. Wohrmann, "Antifreeze glycoprotein in fishes' structure, mode of action and possible applications", *Tierarztliche Praxis Ausgabe Kleintiere Heimtiere*, 1996, vol. 24, no. 1, pp. 1–19.

17 S. Juan, "Interview with a vampire", *National Post* (Toronto), 19 October 2004, "Body & Health", p. 1.

18 S. Juan, "We're all true blue bloods", *New York Daily News*, 8 June 2005, "Body Work", pp. 1–2.

19 J. Travis, "Kids: Getting under mom's skin for decades", *Science News*, 1996, vol. 149, p. 85.

20 S. Juan, "What is a chimera?", *National Post* (Toronto), 18 April 2005, "Body & Health", pp. 1–2.

21 S. Juan, "Getting to grips with the jugular vein", *National Post* (Toronto), 2 May 2005, "Body & Health", p. 2.

22 B. Gobel, "Disseminated intravascular coagulation", *Seminars in Oncological Nursing*, 1999, vol. 15, no. 3, pp. 174–182. Dr Gobel is from the Gottlieb Memorial Hospital in Melrose Park, Illinois.

23 Dr Patrice Guyenet is from the Department of Pharmacology at the University of Virginia in Charlottesville.

24 Dr George Richerson is from the Department of Neurology at the School of Medicine at Yale University.

25 S. Juan, "When the lungs say, 'Get out!'", *New York Daily News*, 15 June 2005, "Body Work", pp. 1–2.

26 S. Juan, "How can Tibetan villagers live at such high altitudes?", *National Post* (Toronto), 31 October 2005, "Body & Health", pp. 1–2.

27 T. Guy, "Evolution and current status of the total artificial heart: The search continues", *ASAIO Journal*, 1998, vol. 44, no. 1, pp. 28–33. Dr T.S. Guy is from the Landstuhl Regional Medical Center in Germany.

28 L. Iannucci, "Brave new bodies: The future is here", *American Health*, December 1995, pp. 68–69.

29 D. Mooney and A. Mikos, "Growing new organs", *Scientific American*, April 1999, pp. 60–65.

30 S. Juan, "Synthetic people", *Sun-Herald* (Sydney), 2 May 1999, "Tempo", p. 16.

CHAPTER 11: THE STOMACH & INTESTINES

1 S. Juan, "Satisfy your appetite for hunger questions", *National Post* (Toronto), 18 April 2005, "Body & Health", p. 1.

2 S. Juan, "That growling in your stomach", *New York Daily News*, 24 May 2005, "Body Work", pp. 2–3.

3 J. Bishop, "Panel to decide what is cause of peptic ulcers", *Wall Street Journal*, 7 February 1994, pp. B2–B3.

4 S. Juan, "What does the small intestine do?", *National Post* (Toronto), 20 June 2005, "Body & Health", pp. 1–2.

5 J. Arnold, "Scientific sleuths track the origins of tapeworms in humans", Agricultural Research Service, US Department of Agriculture, 23 October 2001, p. 1.

6 Dr Martin J. Carey is from the Department of Emergency Medicine at the University of Arkansas for Medical Sciences in Little Rock.

7 S. Juan, "One gonad or two? It's all a matter of balance", *National Post* (Toronto), 29 October 2004, "Body & Health", p. 1.

8 Dr Eric Hollander is a Professor of Psychiatry at Mount Sinai School of Medicine in New York.

9 S. Juan, "What is bigorexia? What is orthorexia?", *National Post* (Toronto), 16 May 2005, "Body & Health", pp. 1–2.

10 M. Catalina Zamora, B. Bota Bonaechea, F. Garcia Sanchez and B. Rios Rial, "Orthorexia nervosa. A new eating behavior disorder?" *Actas Espana Psiquiatria*, 2005, vol. 33, no. 1, pp. 66–68. Dr Catalina Zamora and colleagues are from the Mostoles Hospital in Madrid.

11 S. Juan, "Casting wind to the fates", *New York Daily News*, 28 June 2005, "Body Work", pp. 1–2.

12 Personal communication, 5 September 2005. Dr Ng is from the School of Pharmacy and Medical Sciences at the University of South Australia in Adelaide.

13 B. Barrett, "A long meal", *New Scientist*, 4 October 2003, p. 65.

14 A. Klijn, M. Asselman, M. Vijverberg, P. Dik and T. deJong, "The diameter of the rectum on ultrasonography as a diagnostic tool for constipation in children with dysfunctional voiding", *Journal of Urology*, 2004, vol. 172, no. 5, pt 1, pp. 1986–1988.

15 S. Juan, "How long does it take to digest something completely?", *New York Daily News*, 6 September 2005, "Body Work", pp. 1–2.

16 The research team was based in the Institute of Medical Biochemistry and Genetics at the University of Copenhagen in Denmark.

17 H. Eiberg, I. Berendt and J. Mohr, "Assignment of dominant inherited nocturnal enuresis (ENUR1) to chromosome 13q", *Nature Genetics*, 1995, vol. 10, no. 3, pp. 354–356.

18 H. Eiberg, "Total genome scan analysis in a single extended family for primary nocturnal enuresis: Evidence for a new locus (ENUR3) for primary nocturnal enuresis on chromosome 22q11", *European Journal of Urology*, 1998, vol. 33 (supplement 3), pp. 34–36.

19 S. Juan, "The bed-wetting gene", *Sydney's Child*, August 1999, p. 25.

20 A. Romer and T. Parsons, *The Vetebrate Body* (6th edition), Saunders College Publishing, Philadelphia, 1986, p. 354.

21 Personal communication, 13 August 2004. Dr Shoshkes Reiss is an immunologist in the Department of Biology at New York University.

22 C. Ray, "Uses for the appendix", *New York Times*, 22 December 1998, p. D2.

23 Dr R.A. Wheeler and Dr P.S. Malone are surgeons at the Wessex Regional Centre for Paediatric Surgery at the General Hospital in Southampton, UK.

24 R. Wheeler and P. Malone, "Use of the appendix in reconstructive surgery: A case against incidental appendicectomy", *British Journal of Surgery*, 1991, vol. 78, pp. 1283–1285.

25 R. Luong, K. James and J. Uys, "Dead end", *New Scientist*, 8 February 2003, p. 65.

26 S. Juan, "Is it true that King Tut's penis is missing?", *National Post* (Toronto), 17 October 2005, "Body & Health", pp. 1–2.

27 S. Juan, "Synthetic people", *Sun-Herald* (Sydney), 2 May 1999, "Tempo", p. 16.

28 D. Rohrmann, D. Albrecht, J. Hannappel, R. Gerlach, G. Schwarzkopp and W. Lutzeyer, "Alloplastic replacement of the urinary bladder", *Journal of Urology*, 1996, vol. 156, no. 6, pp. 2094–2097.

29 J. Jaremko and O. Rorstad, "Advances toward the implantable artificial pancreas for treatment of diabetes", *Diabetes Care*, 1998, vol. 21, no. 3, pp. 444–450.

CHAPTER 12: OTHERWISE INSIDE

1 J. Wiese, S. McPherson, M. Odden and G. Shlipak, "Effect of Opuntia ficus indica on symptoms of the alcohol hangover", *Archives of Internal Medicine*, 2004, vol. 164, no. 12, pp. 1334–1340. Dr Wiese and colleagues are from the Tulane Health Sciences Center in New Orleans.

2 S. Juan, "Sobering thoughts", *New York Daily News*, 16 March 2005, "Body Work", p. 1.

3 S. Juan, "Can the cold cause a cold?", *National Post* (Toronto), 14 March 2005, "Body & Health", pp. 1–2.

4 M. Goldwyn, *How a Fly Walks Upside Down . . . And Other Curious Facts*, Wing Books, New York, 1995, p. 236.

5 Dr Allen Rosen is a plastic surgeon from the University of Medicine and Dentistry of New Jersey in Newark and spokesperson for the American Society of Plastic Surgeons.

6   Dr R. Einav-Bachar is from the Institute for Endocrinology and Diabetes at the Schneider Children's Medical Centre of Israel in Petah Tiqva.

7   "Gynecomastia — Abnormal enlargement of the male breast", *Bottom Line*, 1 August 2002, p. 15.

8   R. Einav-Bachar, M. Phillip, Y. Aurbach-Klipper and L. Lazar, "Prepubertal gynaecomastia: Aetiology, course and outcome", *Clinical Endocrinology*, 2004, vol. 61, no. 1, pp. 55–60.

9   Dr Frederick Sweet is from the Department of Obstetrics and Gynecology at the School of Medicine at Washington University in St. Louis.

10 S. Juan, "What makes breasts grow?", *National Post* (Toronto), 2 August 2005, pp. 1–2.

11 S. Juan, "Mind over mammaries", *National Post* (Toronto), 19 March 2005, p. 1.

12 Personal communication, 2 February 2005.

13 S. Juan, "Synthetic people", *Sun-Herald* (Sydney), 2 May 1999, p. 16.

14 L. Iannucci, "Brave new bodies: The future is here", *American Health*, December 1995, pp. 68–69.

15 A. Smith, *The Body*, Viking, New York, 1985.

16 S. Juan, "Water and the body", *National Post* (Toronto), 14 April 2005, "Body & Health", p. 1.

17 S. Juan, "You're all wet", *New York Daily News*, 26 April 2005, "Body Work", p. 1.

18 Dr Brett Ellis is from the Department of Parasitology at Tulane University in New Orleans.

19 Dr Stephan Moorman is from the Department of Anatomy at Case Western Reserve University in Cleveland, Ohio.

20 Y. Kamohara, J. Rozga and A. Demetriou, "Artificial liver: Review and Cedars-Sinai experience", *Journal of Hepatobiliary Pancreatic Surgery*, 1998, vol. 5, no. 3, pp. 273–285.

21 S. Emre, M. Schwartz, D. Likholatnikov, D. Kelly, S. Guy, P. Sheiner and C. Miller, "Split liver transplantation including in-situ split: A single-center experience", paper presented to American Association for the Study of Liver Diseases, 1996.

22 D. Mooney and A. Mikos, "Growing new organs", *Scientific American*, April 1999, pp. 60–65.

23 S. Juan, "Synthetic people", *Sun-Herald* (Sydney), 2 May 1999, "Tempo", p. 16.

CHAPTER 13: SLEEP

1   S. Juan, "While you were sleeping (1)", *Sun-Herald* (Sydney),
    14 November 1999, "Tempo", p. 16.

2   S. Juan, "Can I cause another person to dream?", *New York Daily
    News*, 4 August 2004, "Body Work", pp. 1–2.

3   A. Mavromatis, *Hypnagogia: The Unique State of Consciousness Between
    Wakefulness & Sleep*, Routledge, London, 1987.

4   Personal communication, 17 May 2005.

5   S. Juan, "How long do dreams last?", *National Post* (Toronto), 20 June
    2005, "Body & Health", p. 1.

6   Dr Thomas Balkin is a sleep psychologist at the Walter Reed General
    Hospital in Bethesda, Maryland, and Dr Allen Braun is a
    neuroscientist at the National Institute of Health in Washington, DC.

7   Personal communication, 13 May 2005.

8   Personal communication, 15 May 2005.

9   S. Juan, "Why does my fiancé sleep with one eye open?", *National
    Post* (Toronto), 20 June 2005, "Body & Health", p. 2.

10 D. Rosenfeld and A. Elhajjar, "Sleepsex: A variant of sleepwalking",
    Archives of Sexual Behavior, 1998, vol. 23, no. 3, pp. 269–278. Dr David
    Rosenfeld and Dr Antoine Elhajjar are from the Sleep Medicine
    Clinic Laboratory at Kaiser Permanente Medical Center in Los
    Angeles.

11 M. Mangan, *Sleepsex: Uncovered*, Xlibris, Philadelphia, 2001.

12 C. Guilleminault, A. Moscovitch, K. Yuen and D. Poyares, "Atypical
    sexual behavior during sleep", *Psychosomatic Medicine*, 2002, vol. 64,
    no. 2, pp. 328–336.

13 S. Juan, "The things people do in the night", *National Post* (Toronto),
    11 July 2005, "Body & Health", pp. 1–2.

14 *International Classification of Sleep Disorders, Diagnostic and Coding
    Manual*, American Sleep Disorders Association and American
    Academy of Sleep Medicine, Rochester, Minnesota, 2000.

15 S. Juan, "While you were sleeping (2)", *Sun-Herald* (Sydney), 3
    September 2000, "Tempo", p. 8.

16 S. Juan, "Sleeping easy", *Sun-Herald* (Sydney), 15 October 2000,
    "Tempo", p. 15.

17 T. Vago, "Virginity and defloration", *HaRefuah*, 2000, vol. 139,
    pp. 316–317. Dr Vago is from the Endocrinological and Diabetic
    Services of Milan, Italy.

18 S. Juan, "Is there any truth to the belief that sleeping with a young
    woman rejuvenates an old man?", *National Post* (Toronto),
    17 October 2005, "Body & Health", pp. 1–2.

19   Dr Hui-Ling Lai is Vice Director of Nursing at the Buddhist Tzu
     Chi General Hospital and Assistant Professor at Tzu Chi University
     in Taiwan.

20   H. Lai and M. Good, "Music improves sleep quality in older adults",
     Journal of Advanced Nursing, 2005, vol. 49, no. 3, pp. 234–244.

21   Personal communication, 12 January 1999.

CHAPTER 14: ENDINGS

1   Personal communication, 14 February 1999.

2   Personal communication, 15 February 1999.

3   Personal communication, 31 January 1999.

4   R. Jackson, A. Ramsay, C. Christensen, S. Beaton, D. Hall and
     I. Ramshaw, "Expression of mouse interleukin 4 by a recombinant
     ectromelia virus suppresses cytotoxic lymphocyte responses and
     overcomes genetic resistance to mousepox", *Journal of Virology*, 2000,
     vol. 75, pp. 1205–1210.

5   T. Haga, T. Kuwata, J. Kozyrev, T. Kowfie, M. Hayami and T. Miura,
     "Construction of a SIV/HIV type 1 chimeric virus with the human
     interleukin 6 gene and its production of interleukin 6 in monkey and
     human cells", *Aids Research and Human Retroviruses*, 2000, vol. 16,
     pp. 577–582.

6   S. Juan, "Could researchers accidentally create a virus that wipes out
     all humans?", *National Post* (Toronto), 19 December 2005, "Body &
     Health", pp. 1–2.

7   S. Juan, "How long does it take muscle to die?", *National Post*
     (Toronto), 7 November 2005, "Body & Health", pp. 1–2.

8   J. Downs, "How long does it take for a human body to
     decompose?", *Discover*, July 2002, p. 14.

9   S. Juan, "What is livor mortis?", *National Post* (Toronto), 31 October
     2005, "Body & Health", p. 2.

10   S. Juan, "Why are germs so bad for you?", *National Post* (Toronto), 24
     October 2005, "Body & Health", p. 2.

11   S. Juan, "Lightning strikes", *National Post* (Toronto), 12 September
     2005, "Body & Health", p. 1.

12   Professor Stephen A. Nelson is from the Department of Geological
     and Environmental Sciences at Tulane University in New Orleans.

13   S. Juan, "Did a woman once turn into soap?", *National Post*
     (Toronto), 2 August 2005, "Body & Health", p. 1.

14   K. Iverson, *Death to Dust: What Happens to Dead Bodies?* (2nd
     edition), Galen Press, Tucson, Arizona, 2001.

15   S. Juan, "How long does it take a body to decompose?", *New York
     Daily News*, 12 July 2005, "Body Work", pp. 1–2.

16 Institute of Child Health, Great Ormond Street Hospital for Children, *Peanut Allergy*, Great Ormond Street Hospital for Children NHS Trust, London, 2005.

17 D. Halpern and S. Coren, "Do right-handers live longer?", *Nature*, 1988, vol. 333, p. 213. Dr Diane Halpern is from the California State University at San Bernadino and Dr Stanley Coren is from the University of British Columbia.

18 W. London, "Left-handedness and life expectancy", *Perceptual & Motor Skills*, 1989, vol. 68, no. 3, pp. 1040–1042. Dr London is from Dartmouth Medical School in the UK.

19 M. Anderson, "Lateral preference and longevity", *Nature*, 1989, vol. 341, p. 112. Dr Max Anderson is from the Canadian Statistical Analysis Service in Vancouver.

20 C. Porac and I. Friesen, "Hand preference side and its relation to hand preference switch history among old and oldest-old adults", *Developmental Neuropsychology*, 2000, vol. 17, no. 2, pp. 225–239. Dr Clare Porac is a Professor of Psychology at Pennsylvania State University in Erie and Dr Ingrid Friesen is from the Department of Psychology at the University of Victoria in British Columbia.

21 O. Basso, J. Olsen, N. Holm, A. Skytthe, J. Vaupel and K. Christensen, "Handedness and mortality: A follow-up study of Danish twins born between 1900 and 1910", *Epidemiology*, 2000, vol. 11, no. 5, pp. 576–580.

22 S. Juan, "Why do right-handers live longer than left-handers? If both hands were severed and sewn onto the opposite arm would I still be right-handed or left-handed?" *National Post* (Toronto), 30 May 2005, "Body & Health", pp. 1–2.

23 K. Otterbein, *How War Began*, Texas A&M University Press, College Station, 2004.

24 L. Keeley, *War Before Civilization: The Myth of the Peaceful Savage*, Oxford University Press, New York, 1996. Lawrence H. Keeley is from the University of Illinois.

25 L. deMause, *The Emotional Life of Nations*, Other Press, New York, 2002. Lloyd deMause is head of the Institute for Psychohistory in New York.

26 S. Juan, "What happens when you are executed by hanging?", *National Post* (Toronto), 16 May 2005, "Body & Health", p. 1.

27 World Health Organization, *Leprosy Today*, World Health Organization, Geneva, 2004.

28 S. Juan, "Leprosy need not be a disfiguring disease", *National Post* (Toronto), 16 May 2005, "Body & Health", p. 2.

29 S. Hawking, *A Brief History of Time: From the Big Bang to Black Holes*, Transworld, London, 1988.

30 S. Juan, "Muscles not with the strength", *Sun-Herald* (Sydney), 2 April 1999, "Tempo", p. 17.

31 A. Schrag, Y. Ben-Shlomo and N. Quinn, "Prevalence of progressive supranuclear palsy and multiple system atrophy: A cross-sectional study", *The Lancet*, 1999, vol. 354, no. 9192, pp. 1771–1775. Dr A. Schrag and colleagues are from the Institute of Neurology in London.

32 Personal communication, 19 January 2000.

33 H. Morris, N. Wood and A. Lees, "Progressive supranuclear palsy (Steele–Richardson–Olszewski Disease)", *Postgraduate Medical Journal*, 1999, vol. 75, no. 888, pp. 579–584. Dr Morris and colleagues are from the National Hospital for Neurology and Neurosurgery in London.

34 M. Ezquerra, P. Pastor, F. Valleoriola, J. Molinuevo, R. Blesa, E. Tolosa and R. Oliva, "Identification of a novel polymorphism in the promoter region of the tau gene highly associated to progressive supranuclear palsy in humans", *Neuroscience Letters*, 1999, vol. 27, no. 5, pp. 183–186. Dr Ezquerra and colleagues are from the Faculty of Medicine at the University of Barcelona.

35 C. Ray, "A rare disease", *New York Times*, 2 April 2002, "Q & A", p. D2.

36 S. Juan, "Searching for a cure", *Sun-Herald* (Sydney), 30 January 2000, "Tempo", p. 12.

37 Personal communication, 19 January 2000. Dr Daniel Lago is from the Department of Psychology at Pennsylvania State University.

38 Ontario Veterinary Medical Association, *Pet Owners' Human–Animal Bond*, Ontario Veterinary Medical Association, Milton, Ontario, 2005.

39 S. Juan, "Animal magic", *Sun-Herald* (Sydney), 25 April 1999, "Tempo", p. 20.

40 S. Juan, "Can owning a pet make you live longer?", *National Post* (Toronto), 30 January 2006, "Body & Health", p. 2.

41 Dr Erin Cram is from the Department of Molecular and Cellular Biology at the University of California at Berkeley.

42 D. Philips, "'Scared to death', more than just an expression", Public Affairs Office, University of California, San Diego, 27 December 2001. Dr Phillips is from the Department of Sociology at the University of California in San Diego.

43 *2000 Year Book Australia*, Australian Bureau of Statistics, Canberra, 2000.

INDEX